Uncertain Analysis in Finite Elements Models

Authored by

Wenhui Mo

School of Mechanical Engineering
Hubei University of Automotive Technology
China

Uncertain Analysis in Finite Elements Models

Author: Wenhui Mo

ISBN (Online): 978-981-5079-06-7

ISBN (Print): 978-981-5079-07-4

ISBN (Paperback): 978-981-5079-08-1

Published by Bentham Science Publishers Pte. Ltd. Singapore. All Rights Reserved.

need for a court order if at any point you breach any terms of this License Agreement. In no event will any delay or failure by Bentham Science Publishers in enforcing your compliance with this License Agreement constitute a waiver of any of its rights.

3. You acknowledge that you have read this License Agreement, and agree to be bound by its terms and conditions. To the extent that any other terms and conditions presented on any website of Bentham Science Publishers conflict with, or are inconsistent with, the terms and conditions set out in this License Agreement, you acknowledge that the terms and conditions set out in this License Agreement shall prevail.

Bentham Science Publishers Pte. Ltd.
80 Robinson Road #02-00
Singapore 068898
Singapore
Email: subscriptions@benthamscience.net

CONTENTS

PREFACE i

CHAPTER 1 NONLINEAR STOCHASTIC FINITE ELEMENT METHOD 1
    INTRODUCTION 1
    GENERAL NONLINEAR PROBLEMS 3
        Taylor Expansion Method.. 7
        Perturbation Technology.. 9
        Neumann Stochastic Finite Element 11
        Elastic-plastic Problem 12
    CONCLUDING REMARKS 18
    REFERENCES 18

CHAPTER 2 RELIABILITY CALCULATION OF STOCHASTIC FINITE ELEMENT 23
    INTRODUCTION 23
        Reliability Calculation of Static Problems... 25
        Structural Reliability Calculation of Linear Vibration 35
        Reliability of Nonlinear Structures 43
    CONCLUDING REMARKS 48
    REFERENCES 48

CHAPTER 3 FUZZY RELIABILITY CALCULATION BASED ON STOCHASTIC FINITE ELEMENT 52
    INTRODUCTION 52
        Fuzzy Reliability Calculation of Static Problems Based on Stochastic Finite Element.. 54
        Fuzzy Reliability Calculation of Structures with Linear Vibration.................. 59
        Fuzzy Reliability of Nonlinear Structures 60
        Fuzzy Reliability of Structures with Nonlinear Vibration............................... 60
    CONCLUDING REMARKS 60
    REFERENCES 60

CHAPTER 4 STATIC ANALYSIS OF INTERVAL FINITE ELEMENT 63
    INTRODUCTION 63
        Taylor Expansion for Interval Finite Element... 64
        Interval Finite Element using Neumann Expansion 70
        Interval Finite Element using Sherman-Morrison-Woodbury Expansion 74
        A New Iterative Method (NIM)... 75
    CONCLUDING REMARKS 76
    REFERENCES 77

CHAPTER 5 INTERVAL FINITE ELEMENT FOR LINEAR VIBRATION 79
    INTRODUCTION 79
        Interval Perturbation Finite Element for Linear Vibration.............................. 80
        Interval Neumann finite Element for Linear Vibration 85
        Interval Sherman-Morrison-Woodbury Expansion Finite Element for Linear Vibration... 92
        A New Iterative Method (NIM) 94
    CONCLUDING REMARKS 96
    REFERENCES 96

CHAPTER 6 NONLINEAR INTERVAL FINITE ELEMENT 98
    INTRODUCTION 98
        General Nonlinear Problems... 99
        Elastoplastic Problem 106
        The Homotopy Perturbation Method (MIHPD) 116

    CONCLUDING REMARKS 117
    REFERENCES 117

CHAPTER 7 NONLINEAR VIBRATION ANALYSIS OF INTERVAL FINITE ELEMENT...........120
    INTRODUCTION 120
        Interval Perturbation Finite Element for Nonlinear Vibration 122
        Interval Neumann Finite Element for Nonlinear Vibration....................................... 124
        Interval Taylor Finite Element used for Nonlinear Vibration 126
        Interval Sherman-Morrison-Woodbury Expansion Finite Element 128
        The Homotopy Perturbation Method (MIHPD) 128
    CONCLUDING REMARKS 132
    REFERENCES 132

CHAPTER 8 RANDOM FIELD, INTERVAL FIELD, FUZZY FIELD AND

MIXED FIELD...134
    INTRODUCTION 134
        Stochastic Field...135
        Improved Interpolation Method 136
        Interval Field...138
    CONCLUDING REMARKS 145
    REFERENCES 145

CHAPTER 9 MIXED FINITE ELEMENT 147
    INTRODUCTION 147
        Stochastic and Interval Finite Element 148
        Neumann Expansion Method..152
        Taylor Expansion Method..155
        Neumann Expansion Method..157
    CONCLUDING REMARKS 160
    REFERENCES 160

SUBJECT INDEX 163

PREFACE

There are three kinds of uncertainties in engineering problems. One is randomness, the second is fuzziness, and the third is non probability. Sometimes, the impact of uncertainty on engineering problems can not be ignored. Uncertainty has a great impact on buildings, dams, nuclear power plants, bridges, aircraft, machinery, vehicles, warship, *etc*. The material properties, geometry parameters and loads of the structure are assumed to be random, fuzzy and non probabilistic.

In the first chapter, nonlinear stochastic finite elements for general nonlinear problems and elastoplastic problems are discussed, and three methods are proposed. In Chapter 2, the calculation formula of stochastic finite element is given by using the third-order Taylor expansion and a simple calculation method is addressed. The stress-strength interference model, Monte Carlo simulation, a new iterative method (NIM) of reliability calculation for the linear static problem and linear vibration are proposed. Reliability calculation methods using the homotopy perturbation method (MIHPD) and second order reliability method for the nonlinear static problem and nonlinear vibration are proposed. In Chapter 3, the structural fuzzy reliability calculation of static problem, linear vibration, nonlinear problem and nonlinear vibration is studied by using the stochastic finite element method. The normal membership function is selected as the membership function, and the calculation formula of fuzzy reliability is presented. In Chapter 4, Taylor expansion, Neumann expansion, Sherman Morrison Woodbury expansion and a new iterative method (NIM) for interval finite element calculation of static problems are proposed. In Chapter 5, Perturbation technology, Taylor expansion, Neumann expansion, Sherman Morrison Woodbury expansion and a new iterative method (NIM) for interval finite element calculation of structural linear vibration are addressed. Chapter 6 proposes five calculation methods of nonlinear interval finite element for general nonlinear problems and elastoplastic problems. In the seventh chapter, five methods of interval finite element calculation methods for nonlinear structures are presented. In the eighth chapter, two improved methods of random field are proposed. The midpoint method, local average method, interpolation method and improved interpolation method of interval field and fuzzy field are proposed. The calculation method of mixed field is introduced. In the last chapter, calculation methods of random interval finite element, random fuzzy finite element and random fuzzy and interval finite element are proposed by using Taylor expansion and Neumann expansion.

CONSENT FOR PUBLICATION

Not applicable.

CONFLICT OF INTEREST

The author declares no conflict of interest, financial or otherwise.

ACKNOWLEDGEMENTS

Declared none.

Wenhui Mo
School of Mechanical Engineering
Hubei University of Automotive Technology
China

CHAPTER 1

Nonlinear Stochastic Finite Element Method

Abstract: Considering the influence of random factors on the structure, three stochastic finite element methods for general nonlinear problems are proposed. They are Taylor expansion method, perturbation method and Neumann expansion method. The mean value of displacement is obtained by the tangent stiffness method or the initial stress method of nonlinear finite elements. Nonlinear stochastic finite element is transformed into linear stochastic finite element. The mean values of displacement and stress are obtained by the incremental tangent stiffness method and the initial stress method of the finite element of elastic-plastic problems. The stochastic finite element of elastic- plastic problems can be calculated by the linear stochastic finite element method.

Keywords: Nonlinear stochastic finite element method, Taylor expansion, Perturbation technology, Neumann expansion, Elastic-plastic problem.

INTRODUCTION

In the fields of dams, buildings, earthquakes and so on, random factors have a great impact on the structure. Under random load and working environment, advanced numerical technology and famous finite element method are used to analyze structures. Most applications are limited to certain loads and working environments, although random and uncertain factors reach a considerable degree. Due to the spatial variability of material properties and the randomness of load, the research of stochastic finite elements has attracted more and more attention by many authors. Nonlinear structures are affected by random factors. In order to improve the calculation accuracy, the research of nonlinear stochastic finite elements is very necessary.

The second order perturbation and stochastic second central moment technique solve homogenization of two-phase elastic composites [1]. Both Monte Carlo Simulation and perturbation Methods are examined [2]. The second-order perturbation and the second probabilistic moment method are applied to the stress-

Wenhui Mo

based finite element method [3]. The variability of displacements and eigenvalues of structures are studied using variability response functions [4]. The weighted integral and local average methods of the triangular composite facet shell element are presented [5]. The nonlinear behaviour of strand-based wood composites is simulated [6]. A stochastic formulation of shell structures with multiple uncertain materials and geometric properties is proposed [7]. Efficient approaches to finite element analysis reinforced concrete structures are dealt with [8]. Stochastic finite element analysis of shells is presented for the case of combined uncertain material and geometric properties [9]. A novel response surface method employing ad hoc ratios of polynomials as a performance function is presented [10]. Efficient iterative algorithms for the stochastic finite element method with application to acoustic scattering for solving problems are described [11]. This work studies elliptic boundary value problems with uncertain coefficients by the stochastic finite element method [12]. A projection scheme based on the preconditioned stochastic Krylov subspace is compared with [13]. Primal mixed finite-element approximation of the second-order elliptic problem is proposed [14]. A biodegradation using the perturbation based stochastic finite element method is analyzed [15]. A perturbation-based stochastic finite element method can be applied to solve some boundary values [16]. Multiscale finite element methods are applied to uncertainty quantification [17]. An alternative unsteady adaptive stochastic finite element is proposed to further improve the accuracy [18]. A Galerkin-based multi –point reduced –order model (ROM) is developed for design optimization [19]. The generalized spectral decomposition method for nonlinear stochastic problems is extended [20]. The stochastic perturbation technique and Monte Carlo simulation (MCS) method are used to analyse a cable-stayed bridge system with varying material properties [21]. This paper presents a generic high dimensional model for stochastic finite element analysis [22]. Nonlinear bending response of laminated composite spherical shell panel with random system properties is investigated [23]. The investigation reported an approach for nonlinear multi-degree-of -freedom systems with uncertain properties to non-Gaussian random excitations [24]. This paper presents a stochastic nonlinear failure analysis of laminated composite plates with random material properties [25]. We consider the convergence properties of return algorithms in computational elasto-plasticity [26]. A plastic–damage model of seismic cracking of concrete gravity dams utilizing two different damping mechanisms is examined [27]. This work examines the effect of random geometric imperfections in the buckling analysis of portal frames with stochastic imperfections [28]. This paper investigates the effect of measurement noise and excitation for nonlinear finite element model updating [29]. This paper investigates stochastic analysis of structures with softening materials [30]. Nonlinear finite element modeling of reinforced concrete haunched beams are presented and discussed [31]. This work explores reliability sensitivity analysis of nonlinear structural systems under stochastic excitation [32]. This paper introduces adaptive condensed SFEs for nonlinear mechanical problems [33]. The Markov diffusion theory is applied in studying stochastic nonlinear ships rolling in random beam seas [34]. In this work, a computational framework for nonlinear finite element models is presented [35]. Subset simulation, a Markov chain Monte Carlo technique, can be used to estimate physical models and nonlinear finite element analysis [36]. In this paper, the nonlinear finite element approach is used to solve laminated composite thin hypar shells [37]. This paper investigates randomness in constituent material properties by presenting a reliable model for solving stochastic nonlinear equations [38]. The paper proposes a spectral stochastic formulation with a nonlinear analysis of framed structures [39]. This paper presents a nonlinear finite element model of adobe masonry structures [40]. A nonlinear proxy finite element analysis (PFEA) technique was developed to predict capacity assessment of older t-beam bridges [41]. Stochastic constitutive modeling of elastic-plastic is developed that is efficient for use in which the material properties are considered random variables [42]. To analyze the influence of uncertainty, the methodology of stochastic cohesive interface analysis of layer debonding is proposed [43].

The finite element has become an important method for analyzing structures. Nonlinear structures are affected by random factors and

sometimes they can not be ignored. Three stochastic finite element methods for general nonlinear problems are proposed. Three stochastic finite element methods for elastic-plastic problems are formulated.

GENERAL NONLINEAR PROBLEMS

When the material stress-strain is nonlinear, the stiffness matrix is not constant, which is related to strain and displacement. The global equilibrium equation of the structure is the following nonlinear equations

$$\{\emptyset\} = [A(U)]\{U\} - \{F\} = 0 \tag{1}$$

where $[A(U)]$ is global stiffness matrix, $\{U\}$ is displacement matrix and $\{F\}$ is load matrix. $[A(U)]$ is a system of nonlinear equations.

Many parameters of materials have spatial variabilities, such as elastic modulus and Poisson's ratio. They are assumed to be random processes or random variables. Loads are assumed to be random processes or random variables. Geometry parameters are also assumed to be random variables. Represent the mean value of elastic modulus, Poisson's ratio, geometry parameters and loads into the above equation. The tangent stiffness method is used to solve Eq.1.

The stress-strain relationship of materials is expressed in an incremental form

$$d\{\sigma\} = [D_T(\{\varepsilon\}\,)\,]d\{\varepsilon\} \tag{2}$$

where $[D_T\{\varepsilon\}]$ is tangent elasticity matrix, $d\{\sigma\}$ are incremental stress matrix and $d\{\varepsilon\}$ is incremental strain matrix .

$$\frac{d\{\emptyset\}}{d\{U\}} = [A_T] \tag{3}$$

where $[A_T]$ is the tangent stiffness matrix.

$$[A_T] = \int_V [B]^T [D_T(\{\varepsilon\})][B]dV \tag{4}$$

where $[B]$ is element strain matrix.

The iterative formula is

$$\Delta\{U\}_n = -[A_T]_n^{-1}\{\emptyset\}_n \tag{5}$$

$$\{U\}_{n+1} = \{U\}_n + \Delta\{U\}_n \tag{6}$$

where $\{\emptyset\}_n = \int_V [B]^T \{\sigma\}_n dv - \{F\}$, $\Delta\{U\}_n$ is an increment of displacement matrix in the nth iteration , $\{U\}_n$ is displacement matrix in the nth iteration and $\{U\}_{n+1}$ is displacement matrix in the n+1th iteration.

The initial stress method is used to solve Eq. 1

The stress-strain relationship of a material is expressed as

$$\{\sigma\} = f(\{\varepsilon\}) \tag{7}$$

where $\{\sigma\}$ is stress matrix, $f(\{\varepsilon\})$ is a function of strain

The linear elastic stress-strain relationship with initial stress is

$$\{\sigma\} = [D]\{\varepsilon\} + \{\sigma_0\} \tag{8}$$

where $\{\sigma_0\}$ is initial stress value. $[D]$=the material response matrix of element $d\{\varepsilon\}$ is strain matrix .

Then

$$\{\sigma_0\} = f(\{\varepsilon\}) - [D]\{\varepsilon\} = \{\sigma\} - \{\sigma\}_e \tag{9}$$

Where

$$\{\sigma\}_e = [D]\{\varepsilon\} \tag{10}$$

Let

$$[A_T]_0 = \int_V [B]^T [D][B] dV \tag{11}$$

where $[A_T]_0$ is the initial tangent stiffness matrix of the structure.

Substituting the Eq. 11 into Eq. 1, we obtain

$$[A_T]_0\{U\} = \{F\} + \{\overline{F}\} \tag{12}$$

Where

$$\{\overline{F}\} = -\int_V [B]^T \{\sigma_0\} dV \tag{13}$$

The first approximate displacement is calculated by the following formula

$$\{U\}_1 = [A_T]_0^{-1}\{F\} \tag{14}$$

where $\{U\}_1$ is the first approximate displacement.

The initial stress is

$$\{\sigma_0\}_1 = f(\{\varepsilon\}_1) - [D]\{\varepsilon\}_1 \tag{15}$$

We obtain

$$\{\overline{F}\}_1 = -\int_V [B]^T \{\sigma_0\}_1 dV \tag{16}$$

where $\{\overline{F}\}_1$ is the joint load that is equivalent to the initial stress.

$$\Delta\{U\}_1 = [A_0]^{-1}\{\overline{F}\}_1 \tag{17}$$

where $\Delta\{U\}_1$ is the increment of displacement after the first calculation.

$${U}_2 = {U}_1 + \Delta{U}_1 \tag{18}$$

where $\{U\}_2$ is the second approximation of the displacement after the first calculation.

The iterative equation is expressed as

$${U}_{n+1} = {U}_n + \Delta{U}_n \tag{19}$$

where $\{U\}_n$ is the approximate value of the displacement after the nth calculation. $\{U\}_{n+1}$ is the approximate value of the displacement after the n+1th calculation. $\Delta\{U\}_n$ is the increment of displacement after the nth calculation.

After $\{U\}_{n+1}$ represents Eq.1, it becomes a linear equation containing stochastic variables of Young's modulus , Poisson's ratio ,geometry parameters and loads. Nonlinear stochastic finite element is transformed into linear stochastic finite element.

Taylor Expansion Method

Eq.1 is rewritten as

$$[A]{U} = {F} \tag{20}$$

The elastic modulus , Poisson's ratio ,geometry parameters and loads of the structure are assumed to be n random variables $a_1, a_2, \cdots, a_i, \cdots, a_n$.

The partial derivative of Eq.20 with respect to a_i is given by

$$\frac{\partial \{U\}}{\partial a_i} = [A]^{-1} \left(\frac{\partial \{F\}}{\partial a_i} - \frac{\partial [A]}{\partial a_i} \{U\} \right) \tag{21}$$

where $\dfrac{\partial \{U\}}{\partial a_i}$ is the partial derivative of $\{U\}$ with respect to a_i.

The partial derivative of Eq.21 with respect to a_j is given by

$$\frac{\partial^2 \{U\}}{\partial a_i \partial a_j} = [A]^{-1} \left(\frac{\partial^2 \{F\}}{\partial a_i \partial a_j} - \frac{\partial [A]}{\partial a_i} \frac{\partial \{U\}}{\partial a_j} - \frac{\partial [A]}{\partial a_j} \frac{\partial \{U\}}{\partial a_i} - \frac{\partial^2 [A]}{\partial a_i \partial a_j} \{U\} \right) \tag{22}$$

where $\dfrac{\partial^2 \{U\}}{\partial a_i \partial a_j}$ is the partial derivative of $\dfrac{\partial \{U\}}{\partial a_i}$ with respect to a_j.

The displacement is expanded at the mean point $\bar{a} = (\bar{a}_1, \bar{a}_2, \cdots, \bar{a}_n)^T$ of the random variables, and the mean value is taken on both sides, we obtain

$$E\{U\} \approx \{U\}\big|_{a=\bar{a}} + \frac{1}{2} \sum_{i=1}^{n} \sum_{j=1}^{n} \frac{\partial^2 \{U\}}{\partial a_i \partial a_j}\bigg|_{a=\bar{a}} Cov\left(a_i, a_j\right) \tag{23}$$

where $E\{U\}$ is the mean value of $\{U\}$ and $Cov(a_i, a_j)$ is the covariance of a_i and a_j.

The covariance of any two components of the displacement is

$$Cov\left[\{U\}_{i_1}, \{U\}_{j_1}\right] \approx \sum_{i=1}^{n}\sum_{j=1}^{n} \frac{\partial\{U\}_{i_1}}{\partial a_i}\bigg|_{a=\bar{a}} \cdot \frac{\partial\{U\}_{j_1}}{\partial a_j}\bigg|_{a=\bar{a}} \cdot Cov(a_i, a_j) \qquad (24)$$

where $Cov\left[\{U\}_{i_1}, \{U\}_{j_1}\right]$ is the covariance of any two components of the displacement. $\dfrac{\partial\{U\}_{i_1}}{\partial a_i}\bigg|_{a=\bar{a}}$ is the partial derivative value at the mean point of the random variables. $\dfrac{\partial\{U\}_{j_1}}{\partial a_j}\bigg|_{a=\bar{a}}$ is the partial derivative value at the mean point of the random variables.

Perturbation Technology

See the above section for the mathematical treatment of transfor- ming nonlinear equations into linear equations.

$[A], \{U\}$ and F are expanded at the mean point $\bar{a} = \left(\bar{a}_1, \bar{a}_2, \cdots, \bar{a}_n\right)^T$ of the random variables *via* Taylor series

$$A = A^0 + \sum_{i=1}^{n} \frac{\partial A}{\partial a_i}\alpha_i + \sum_{i=1}^{n} \frac{\partial^2 A}{\partial a_i^2}\alpha_i^2 + \cdots \qquad (25)$$

$$U = U^0 + \sum_{i=1}^{n} \frac{\partial U}{\partial a_i} \alpha_i + \sum_{i=1}^{n} \frac{\partial^2 U}{\partial a_i^2} \alpha_i^2 + \cdots \tag{26}$$

$$F = F^0 + \sum_{i=1}^{n} \frac{\partial F}{\partial a_i} \alpha_i + \sum_{i=1}^{n} \frac{\partial^2 F}{\partial a_i^2} \alpha_i^2 + \cdots \tag{27}$$

By applying perturbation technique, the following equations are given by

$$U^0 = A^{0^{-1}} F^0 \tag{28}$$

$$\frac{\partial \{U\}}{\partial a_i} = [A]^{-1} \left(\frac{\partial \{F\}}{\partial a_i} - \frac{\partial [A]}{\partial a_i} \{U\} \right) \tag{29}$$

$$\frac{\partial^2 \{U\}}{\partial a_i \partial a_j} = [A]^{-1} \left(\frac{\partial^2 \{F\}}{\partial a_i \partial a_j} - \frac{\partial [A]}{\partial a_i} \frac{\partial \{U\}}{\partial a_j} - \frac{\partial [A]}{\partial a_j} \frac{\partial \{U\}}{\partial a_i} - \frac{\partial^2 [A]}{\partial a_i \partial a_j} \{U\} \right) \tag{30}$$

The mean of the displacement is obtained as

$$E\{U\} \approx \{U\}\big|_{a=\bar{a}} + \frac{1}{2} \sum_{i=1}^{n} \sum_{j=1}^{n} \frac{\partial^2 \{U\}}{\partial a_i \partial a_j} \bigg|_{a=\bar{a}} Cov(a_i, a_j) \tag{31}$$

where $E\{U\}$ is the mean value of $\{U\}$ and $Cov(a_i, a_j)$ is the covariance of a_i and a_j.

The covariance of any two components of the displacement is

$$Cov\left[\{U\}_{i_1}, \{U\}_{j_1} \right] \approx \sum_{i=1}^{n} \sum_{j=1}^{n} \frac{\partial \{U\}_{i_1}}{\partial a_i} \bigg|_{a=\bar{a}} \cdot \frac{\partial \{U\}_{j_1}}{\partial a_j} \bigg|_{a=\bar{a}} \cdot Cov(a_i, a_j) \tag{32}$$

where $Cov\left[\{U\}_{i_1}, \{U\}_{j_1}\right]$ is the covariance of any two components of the displacement. $\dfrac{\partial \{U\}_{i_1}}{\partial a_i}\Big|_{a=\bar{a}}$ is the partial derivative value at the mean point of the random variables. $\dfrac{\partial \{U\}_{j_1}}{\partial a_j}\Big|_{a=\bar{a}}$ is the partial derivative value at the mean point of the random variables.

Neumann Stochastic Finite Element

See the above section for the mathematical treatment of transfor- mating nonlinear equations into linear equations.

The stiffness matrix A is decomposed into two matrices

$$[A] = [A_0] + [\Delta A] \tag{33}$$

where A_0 is the stiffness matrix is replaced by mean values, ΔA representing deviatoric parts. The solution U_0 can be obtained as

$$U_0 = A_0^{-1} F \tag{34}$$

The Neumann expansion of A^{-1} takes the following form:

$$[A]^{-1} = ([A_0] + \Delta[A])^{-1} = (I - P + P^2 - P^3 + \cdots)[A_0]^{-1} \tag{35}$$

This series solution is equivalent to the following equation [44]:

$$[A_0]\{U\}_{(i)} = [\Delta A]\{U\}_{(i-1)}, \quad i = 1, 2, \cdots \tag{36}$$

The mean of $\{U\}$ is given by

$$\mu\{U\} = \frac{\{U\}_1 + \{U\}_2 + \cdots + \{U\}_{N_1}}{N_1} \tag{37}$$

where $\mu\{U\}$ is the mean of $\{U\}$

The variance of $\{U\}$ is given by

$$Var\{U\} = \frac{1}{N_1 - 1} \sum_{i=1}^{N_1} \left(\{U\}_i - \mu\{U\} \right)^2 \tag{38}$$

where $Var\{U\}$ is the variance of $\{U\}$.

Elastic-plastic Problem

Solution of displacement

$$\{\varnothing\} = [A(U)]\{U\} - \{F\} = 0 \tag{39}$$

Many parameters of materials have spatial variabilities, such as elastic modulus and Poisson's ratio. They are assumed to be random processes or random variables. Loads are assumed to be random processes or random variables. Geometry parameters are also assumed to be random variables. Represent the mean value of elastic modulus, Poisson's ratio ,geometry parameters and loads into the above equation. The incremental tangent stiffness method is used to solve Eq. 39

The first choice is to conduct a linear elastic analysis to obtain the displacement, strain and stress under elastic load, which are respectively recorded as $\{U\}_0, \{\varepsilon\}_0, \{\sigma\}_0$. On this basis, the load is divided into n increments and acted on the structure successively.

When the load increment $\Delta\{F\}_1$ is applied, the elastic stiffness matrix of the element is

$$[A] = \int_V [B]^T [D]_e [B] dV \tag{40}$$

where $[D]_e$ is elastic matrix.

In the plastic element, the element stiffness matrix is

$$[A] = \int_V [B]^T [D]_{ep} [B] dV \tag{41}$$

Where $[D]_{ep}$ is elastic-plastic matrix.

Solving

$$\Delta\{U\}_1 = [A(\{\sigma\}_0)]^{-1} \Delta\{F\}_1 \tag{42}$$

We can obtain $\Delta\{U\}_1$, and thus obtain $\Delta\{\varepsilon\}_1$, $\Delta\{\sigma\}_1$.

$$\{U\}_1 = \{U\}_0 + \Delta\{U\}_1 \tag{43}$$

$$\{\varepsilon\}_1 = \{\varepsilon\}_0 + \Delta\{\varepsilon\}_1 \tag{44}$$

$$\{\sigma\}_1 = \{\sigma\}_0 + \Delta\{\sigma\}_1 \tag{45}$$

where $\Delta\{U\}_1$, $\Delta\{\varepsilon\}_1$ and $\Delta\{\sigma\}_1$ are increments of displace- ment ,strain and stress after first incremental loading. $\{U\}_1$, $\{\varepsilon\}_1$ and $\{\sigma\}_1$ are displacement, strain and stress after first incremental loading.

After the action of the i-th load increment $\Delta\{F\}_i$, we obtain

$$\Delta\{U\}_i=[A(\{\sigma\}_{i-1})]^{-1}\Delta\{F\}_i \tag{46}$$

And then we get

$${U}_i={U}_{i-1} + \Delta\{U\}_i \tag{47}$$

$${\varepsilon}_i={\varepsilon}_{i-1} + \Delta\{\varepsilon\}_i \tag{48}$$

$${\sigma}_i={\sigma}_{i-1} + \Delta\{\sigma\}_i \tag{49}$$

where $\Delta\{U\}_i$, $\Delta\{\varepsilon\}_i$ and $\Delta\{\sigma\}_i$ are increments of displacement, strain and stress after i-th incremental loading. $\{U\}_i$, $\{\varepsilon\}_i$ and $\{\sigma\}_i$ are displacement, strain and stress after i-th incremental loading.

The following initial stress method can also be used

For elastoplastic problems, the incremental form of stress-strain is

$$d\{\sigma\} = [D]_e d\{\varepsilon\}+d\{\sigma_0\} \tag{50}$$

where $[D]_e$ is elastic matrix.

$$d\{\sigma_0\}=-[D]_p d\{\varepsilon\} \tag{51}$$

where $[D]_p$ is plastic matrix.

The increment is used to replace the differential for each loading after the element begins to be plastic.

$$\Delta\{\sigma\} = [D]_e \Delta\{\varepsilon\} + \Delta\{\sigma_0\} \tag{52}$$

$$\Delta\{\sigma_0\} = -[D]_p \Delta\{\varepsilon\} \tag{53}$$

where $\Delta\{\sigma\}$ is an increment of stress matrix , $\Delta\{\sigma_0\}$ is stress increment after one elastic analysis.

The equilibrium equation of displacement increment is

$$[K_0]\Delta\{U\} = \Delta\{F\} + \{\overline{F}(\Delta\{\varepsilon\})\} \tag{54}$$

Where

$$\{\overline{F}(\Delta\{\varepsilon\})\} = \int B^T D_p \Delta\{\varepsilon\} dV \tag{55}$$

The iterative equation of the ith incremental load is

$$\Delta\{U\}_i^{j+1} = [A_0]^{-1}\left(\Delta\{F\}_i + \{\overline{F}\}_i^j\right) \; (j = 0, 1, 2, \cdots) \tag{56}$$

The first iteration is at $\Delta\{U\}_0 = \Delta\{\varepsilon\}_0 = \Delta\{\sigma\}_0 = 0$. The following iteration is based on the previous iteration and loads $\{\overline{F}\}_i^j$. $\Delta\{U\}_i^{j+1}$ is calculated according to the above formula until convergence.

The mean values of elastic modulus, Poisson's ratio, geometry parameters and load are substituted into equation 39, and the mean values of displacement and stress are obtained by using the incremental tangent stiffness method or the initial stress method. The mean value of displacement is substituted into Eq. 39, which contains only random variables, such as elastic modulus, Poisson's ratiogeometry parameters and loads. The mean value of stress is substituted into the following equation (Eq.57), which contains only random variables, such as elastic modulus, Poisson's ratio and geometry parameters. The nonlinear stochastic finite element is transformed into a linear stochastic finite element.The calculation of the mean and covariance of the displacement is the same as that in section 2 and will not be repeated. The mean value and covariance of stress are calculated as

$$\{\sigma\} = [D]_{ep}[B]\{U\} \tag{57}$$

where $[\mathbf{D}]_{\mathbf{ep}}$ is elastoplastic matrix.

The stress for an element d is given by

$$\{\sigma\} = [D]_{ep}[B]\{U\}^d \tag{58}$$

Where $[B]$= element strain matrix of element d and U^d =the element d nodal displacement vector.

The partial derivative of Eq.58 with respect to a_i is given by

$$\frac{\partial\{\sigma\}}{\partial a_i} = \frac{\partial[D]_{ep}}{\partial a_i}[B]\{U\}^d + [D]_{ep}\frac{\partial[B]}{\partial a_i}\{U\}^d + [D]_{ep}[B]\frac{\partial\{U\}^d}{\partial a_i} \tag{59}$$

The partial derivative of Eq.59 with respect to a_j is given by

$$\frac{\partial^2\{\sigma\}}{\partial a_i \partial a_j} = \frac{\partial^2[D]_{ep}}{\partial a_i \partial a_j}[B]\{U\}^d + \frac{\partial[D]_{ep}}{\partial a_i}\frac{\partial[B]}{\partial a_j}\{U\}^d + \frac{\partial[D]_{ep}}{\partial a_i}[B]\frac{\partial\{U\}^d}{\partial a_j}$$

$$+\frac{\partial[D]_{ep}}{\partial a_j}\frac{\partial[B]}{\partial a_i}\{U\}^d + [D]_{ep}\frac{\partial^2[B]}{\partial a_i \partial a_j}\{U\}^d + [D]_{ep}\frac{\partial[B]}{\partial a_i}\frac{\partial\{U\}^d}{\partial a_j}$$

$$+\frac{\partial[D]_{ep}}{\partial a_j}[B]\frac{\partial\{U\}^d}{\partial a_i} + [D]_{ep}\frac{\partial[B]}{\partial a_j}\frac{\partial\{U\}^d}{\partial a_i} + [D]_{ep}[B]\frac{\partial^2\{U\}^d}{\partial a_i \partial a_j} \tag{60}$$

The stress is expanded at the mean value point $\bar{a} = (\bar{a}_1, \bar{a}_2, \cdots, \bar{a}_i, \cdots, \bar{a}_N)^T$ by means of a Taylor series. By taking the expectation operator for two sides of the above Eq.58, the mean of stress is obtained as

$$\mu\{\sigma\} \approx \{\sigma\}|_{a=\bar{a}} + \frac{1}{2}\sum_{i=1}^{N}\sum_{j=1}^{N}\frac{\partial^2\{\sigma\}}{\partial a_i \partial a_j}\Big|_{a=\bar{a}} Cov(a_i, a_j) \tag{61}$$

where $\mu\{\sigma\}$ is the mean value of $\{\sigma\}$ and $Cov(a_i, a_j)$ is the covariance of a_i and a_j

The covariance of σ is given by

$$Cov(\{\sigma\}_{i_2}, \{\sigma\}_{j_2}) \approx \sum_{i=1}^{N}\sum_{j=1}^{N}\frac{\partial\{\sigma\}_{i_2}}{\partial a_i}\Big|_{a=\bar{a}} \cdot \frac{\partial\{\sigma\}_{j_2}}{\partial a_j}\Big|_{a=\bar{a}} \cdot Cov(a_i, a_j) \tag{62}$$

where $Cov(\{\sigma\}_{i_2}, \{\sigma\}_{j_2})$ is the covariance of any two components of the stress, $\frac{\partial\{\sigma\}_{i_2}}{\partial a_i}\Big|_{a=\bar{a}}$ is the partial derivative value at the mean point of the random variables and $\frac{\partial\{\sigma\}_{j_2}}{\partial a_j}\Big|_{a=\bar{a}}$ is the partial derivative value at the mean point of the random variables.

After the mean of $\{U\}$ and the variance of $\{U\}$ calculated by Neumann expansion stochastic finite element method , the mean value and covariance of stress are calculated as follows

Substituting N_1 samples of vector $\bar{a}$ into Eq.58, vectors $\{\sigma\}_1, \{\sigma\}_2, \cdots, \{\sigma\}_{N_1}$ can be obtained.

The mean of $\{\sigma\}$ is given by

$$\mu\{\sigma\} = \frac{\{\sigma\}_1 + \{\sigma\}_2 + \cdots + \{\sigma\}_{N_1}}{N_1} \tag{63}$$

where $\mu\{\sigma\}$ is the mean value of stress.

The variance of $\{\sigma\}$ is given by

$$Var\{\sigma\} = \frac{1}{N_1 - 1} \sum_{i=1}^{N_1} \left(\{\sigma\}_i - \mu\{\sigma\} \right)^2 \tag{64}$$

where $Var\{\sigma\}$ is the variance of stress.

CONCLUDING REMARKS

Random factors sometimes have a great influence on the structure. With the deepening of human understanding, it is not practical to ignore the randomness of the finite element. The nonlinear stochastic finite element is transformed into a linear stochastic finite element for calculation. The stochastic finite element method for general nonlinear problems is presented. The stochastic finite element analysis method for elastoplastic problems is proposed.

REFERENCES

[1] M. Kamiński, and M. Kleiber, "Perturbation based stochastic finite element method for homogenization of two-phase elastic composites", *Comput. Struc.,* vol. 78, no. 6, pp. 811-826, 2000.

http://dx.doi.org/10.1016/S0045-7949(00)00116-4

[2] S. Beatriz, "L.P.de Lima, Nelson F.F.Ebecken, "A comparison of models for uncertainty analysis by the finite element method", *Finite Elem. Anal. Des.,* vol. 34, no. 2, pp. 211-232, 2000.

http://dx.doi.org/10.1016/S0168-874X(99)00039-6

[3] Marcin Kaminski, *Stochastic second-order perturbation approach to the stress-based finite element method,* .

http://dx.doi.org/10.1016/S0020-7683(00)00234-1

[4] L.L. Graham, and G. Deodatis, "Response and eigenvalue analysis of stochastic finite element systems with multiple correlated material and geometric properties", *Probab. Eng. Mech.,* vol. 16, no. 1, pp. 11-29, 2001.

http://dx.doi.org/10.1016/S0266-8920(00)00003-5

[5] J. Argyris, M. Papadrakakis, and G. Stefanou, "finite element analysis of shells", *Comput. Methods Appl. Mech. Eng.,* vol. 191, no. 41-42, pp. 4781-4804, 2002.

http://dx.doi.org/10.1016/S0045-7825(02)00404-8

[6] P.L. Clouston, and F. Lam, "FrankLam, "A stochastic plasticity approach to strength modeling of strand-based wood composites", *Compos. Sci. Technol.,* vol. 62, no. 10-11, pp. 1381-1395, 2002.

http://dx.doi.org/10.1016/S0266-3538(02)00086-6

[7] G. Stefanou, and M. Papadrakakis,

[8] L. Davenne, F. Ragueneau, J. Mazars, and A. Ibrahimbegovic, "Efficient approaches to finite element analysis in earthquake engineering", *Comput. Struc.,* vol. 81, no. 12, pp. 1223-1239, 2003.

http://dx.doi.org/10.1016/S0045-7949(03)00038-5

[9] G. Stefanou, "Stochastic finite element analysis of shells with combined random material and geometric properties", *Computer Methods in Applied Mechanics and Engineering.,* vol. 193, pp. 139-160, 2004.

[10] G. Falsone, and N. Impollonia, "About the accuracy of a novel response surface method for the analysis of finite element modeled uncertain structures", *Probab. Eng. Mech.,* vol. 19, no. 1-2, pp. 53-63, 2004.

http://dx.doi.org/10.1016/j.probengmech.2003.11.005

[11] H.C. Elman, O.G. Ernst, D.P. O'Leary, and M. Stewart, "Efficient iterative algorithms for the stochastic finite element method with application to acoustic scattering", *Comput. Methods Appl. Mech. Eng.,* vol. 194, no. 9-11, pp. 1037-1055, 2005.

http://dx.doi.org/10.1016/j.cma.2004.06.028

[12] I. Babuška, R. Tempone, and G.E. Zouraris, RaúlTempone, Georgios E.Zouraris, "Solving elliptic boundary value problems with uncertain coefficients by the finite element method: the stochastic formulation", *Comput. Methods Appl. Mech. Eng.,* vol. 194, no. 12-16, pp. 1251-1294, 2005.

http://dx.doi.org/10.1016/j.cma.2004.02.026

[13] S.K. Sachdeva, P.B. Nair, and A.J. Keane, "Comparative study of projection schemes for stochastic finite element analysis", *Comput. Methods Appl. Mech. Eng.,* vol. 195, no. 19-22, pp. 2371-2392, 2006.

http://dx.doi.org/10.1016/j.cma.2005.05.010

[14] D. Kim, and E-J. Park, "Eun-JaePark, "Primal mixed finite-element approximation of elliptic equations with gradient nonlinearities", *Comput. Math. Appl.,* vol. 51, no. 5, pp. 793-804, 2006.

http://dx.doi.org/10.1016/j.camwa.2006.03.006

[15] A. Chaudhuri, and M. Sekhar, "Analysis of biodegradation in a 3-D heterogeneous porous medium using nonlinear stochastic finite element method", *Adv. Water Resour.,* vol. 30, no. 3, pp. 589-605, 2007.

http://dx.doi.org/10.1016/j.advwatres.2006.04.001

[16] M. Kamiński, "Generalized perturbation-based stochastic finite elemen tmethodin elastostatics", *Comput. Struc.,* vol. 85, no. 10, pp. 586-594, 2007.

http://dx.doi.org/10.1016/j.compstruc.2006.08.077

[17] P. Dostert, Y. Efendiev, and T.Y. Hou, "Multiscale finite element methods for stochastic porous media flow equations and application to uncertainty quantification", *Comput. Methods Appl. Mech. Eng.,* vol. 197, no. 43-44, pp. 3445-3455, 2008.

http://dx.doi.org/10.1016/j.cma.2008.02.030

[18] A.S. Jeroen, "Witteveen,HesterBijl, "An alternative unsteady adaptive stochastic finite elements formulation based on interpolation at constant phase", *Comput. Methods Appl. Mech. Eng.,* vol. 198, no. 3-4, pp. 578-591, 2008.

http://dx.doi.org/10.1016/j.cma.2008.09.005

[19] K. Maute, G. Weickum, and M. Eldred, "GaryWeickum,MikeEldred, "A reduced-order stochastic finite element approach for design optimization under uncertainty", *Struct. Saf.,* vol. 31, no. 6, pp. 450-459, 2009.

http://dx.doi.org/10.1016/j.strusafe.2009.06.004

[20] A. Nouy, and O.P. Le Maître, "Olivier P.Le Maî^tre, "Generalized spectral decomposition for stochastic nonlinear problems", *J. Comput. Phys.,* vol. 228, no. 1, pp. 202-235, 2009.

http://dx.doi.org/10.1016/j.jcp.2008.09.010

[21] AlemdarBayraktar,SüleymanAdanur, "Stochastic finite element of a cable-stayed bridge system with varying material properties", *Probab. Eng. Mech.,* vol. 25, pp. 279-289, 2010.

http://dx.doi.org/10.1016/j.probengmech.2010.01.008

[22] R. Chowdhury, and S. Adhikari, "High dimensional model representation for stochastic finite element analysis", *Appl. Math. Model.,* vol. 34, no. 12, pp. 3917-3932, 2010.

http://dx.doi.org/10.1016/j.apm.2010.04.004

[23] A. Lal, B.N. Singh, and S. Anand, "B.N.Singh,SohamAnand, "Nonlinear bending response of laminated composite spherical shell panel with system randomness subjected to hygro-thermo-mechanical loading", *Int. J. Mech. Sci.,* vol. 53, no. 10, pp. 855-866, 2011.

http://dx.doi.org/10.1016/j.ijmecsci.2011.07.008

[24] W. Cho, "SolomonTo,"Response analysis of nonlinear multi-degree-of-freedom systems with uncertain properties to non-Gaussian random excitations", *Probab. Eng. Mech.,* vol. 27, no. 1, pp. 75-81, 2012.

http://dx.doi.org/10.1016/j.probengmech.2011.05.010

[25] A. Lal, B.N. Singh, and D. Patel, "B.N.Singh,DipanPatel, "Stochastic nonlinear failure analysis of laminated composite plates under compressive transverse loading", *Compos. Struct.,* vol. 94, no. 3, pp. 1211-1223, 2012.

http://dx.doi.org/10.1016/j.compstruct.2011.11.018

[26] M. Sauter, and C. Wieners, "On the superlinear convergence in computational elasto-plasticity", *Comput. Methods Appl. Mech. Eng.,* vol. 200, no. 49-52, pp. 3646-3658, 2011.

http://dx.doi.org/10.1016/j.cma.2011.08.011

[27] O. Omidi, S. Valliappan, and V. Lotfi, "Seismic cracking of concrete gravity dams by plastic–damage model using different damping mechanisms", *Finite Elem. Anal. Des.,* vol. 63, pp. 80-97, 2013.

http://dx.doi.org/10.1016/j.finel.2012.08.008

[28] V. Papadopoulos, G. Soimiris, and M. Papadrakakis, "Buckling analysis of I-section portal frames with stochastic imperfections", *Eng. Struct.,* vol. 47, pp. 54-66, 2013.

http://dx.doi.org/10.1016/j.engstruct.2012.09.009

[29] S.G. Shahidi, and S.N. Pakzad, "Effect of measurement noise and excitation on Generalized Response Surface Model Updating", *Eng. Struct.,* vol. 75, pp. 51-62, 2014.

http://dx.doi.org/10.1016/j.engstruct.2014.05.033

[30] M. Georgioudakis, G. Stefanou, and M. Papadrakakis, "Stochastic failure analysis of structures with softening materials", *Eng. Struct.,* vol. 61, pp. 13-21, 2014.

http://dx.doi.org/10.1016/j.engstruct.2014.01.002

[31] E.A. Godínez-Domínguez, A. Tena-Colunga, and G. Juárez-Luna, "Arturo Tena-Colunga,GelacioJuárez-Luna, "Nonlinear finite element modeling of reinforced concrete haunched beams designed to develop a shear failure", *Eng. Struct.,* vol. 105, pp. 99-122, 2015.

http://dx.doi.org/10.1016/j.engstruct.2015.09.023

[32] H.A. Jensen, F. Mayorga, and M.A. Valdebenito, "Reliability sensitivity estimation of nonlinear structural systems under stochastic excitation: A simulation-based approach", *Comput. Methods Appl. Mech. Eng.,* vol. 289, pp. 1-23, 2015.

http://dx.doi.org/10.1016/j.cma.2015.01.012

[33] A. Llau, J. Baroth, L. Jason, and F. Dufour, "JulienBaroth,LudovicJason,FrédéricDufour, "Condensed SFEs for nonlinear mechanical problems", *Comput. Methods Appl. Mech. Eng.,* vol. 309, pp. 434-452, 2016.

http://dx.doi.org/10.1016/j.cma.2016.06.014

[34] W. Chai, A. Naess, and B.J. Leira, "Stochastic nonlinear ship rolling in random beam seas by the path integration method", *Probab. Eng. Mech.,* vol. 44, pp. 43-52, 2016.

http://dx.doi.org/10.1016/j.probengmech.2015.10.002

[35] D. Giagopoulos, and A. Arailopoulos, "Computational framework for model updating of large scale linear and nonlinear finite element models using state of the art evolution strategy", *Comput. Struc.,* vol. 192, pp. 210-232, 2017.

http://dx.doi.org/10.1016/j.compstruc.2017.07.004

[36] K.E. David, "Green, "Efficient Markov Chain Monte Carlo for combined Subset Simulation and nonlinear finite element analysis", *Comput. Methods Appl. Mech. Eng.,* vol. 313, pp. 337-361, 2017.

http://dx.doi.org/10.1016/j.cma.2016.10.012

[37] A. Ghosh, and D. Chakravorty, "First ply failure analysis of laminated composite thin hypar shells using nonlinear finite element approach", *Thin-walled Struct.,* vol. 131, pp. 736-745, 2018.

http://dx.doi.org/10.1016/j.tws.2018.07.046

[38] M. Mohammadi, M. Eghtesad, and H. Mohammadi, "Stochastic analysis of pull-in instability of geometrically nonlinear size-dependent FGM micro beams with random material properties", *Compos. Struct.,* vol. 200, pp. 466-479, 2018.

http://dx.doi.org/10.1016/j.compstruct.2018.05.089

[39] V. Papadopoulos, I. Kalogeris, and D.G. Giovanis, "A spectral stochastic formulation for nonlinear framed structures", *Probab. Eng. Mech.,* vol. 55, pp. 90-101, 2019.

http://dx.doi.org/10.1016/j.probengmech.2018.11.002

[40] F. Parisi, C. Balestrieri, and H. Varum, "Nonlinear finite element model for traditional adobe masonry", *Constr. Build. Mater.,* vol. 223, pp. 450-462, 2019.

http://dx.doi.org/10.1016/j.conbuildmat.2019.07.001

[41] A.P. Schanck, and W.G. Davids, "Capacity assessment of older t-beam bridges by nonlinear proxy finite-element analysis", *Structures,* vol. 23, pp. 267-278, 2020.

http://dx.doi.org/10.1016/j.istruc.2019.09.012

[42] Maxime Lacour, "Stochastic constitutive modeling of elastic-plastic materials with uncertain properties ", *Computers and Geotechnics,* vol. 125, 2020.

[43] N. Malkiel, "Stochastic cohesive interface analysis of layer debonding", *International Journal of Solids and Structures,* vol. 226, 2021.

[44] F. Yamazaki, M. Shinozuka, and G. Dasgupta, "Neumann expansion for stochastic finite element analysis", *J. Eng. Mech.,* vol. 114, no. 8, pp. 1335-1354, 1988.

http://dx.doi.org/10.1061/(ASCE)0733-9399(1988)114:8(1335)

[45] S. Dey, T. Mukhopadhyay, and S. Adhikari, *Uncertainty quantification in laminated composites: A meta-model based approach.,* CRC Press, 2018.

http://dx.doi.org/10.1201/9781315155593

Uncertain Analysis in Finite Elements, 2022, 23-51

CHAPTER 2

Reliability Calculation of Stochastic Finite Element

Abstract: The stochastic finite element third-order perturbation method for linear static problems is formulated. The stress-strength interference model, Monte Carlo simulation and a new iterative method (NIM) of reliability calculation for the linear static problem and linear vibration are proposed. Reliability calculation methods using modified iteration formulas by the homotopy perturbation method (MIHPD) and second- order reliability method for a nonlinear static problem and nonlinear vibration are proposed.

Keywords: The third-order Taylor expansion, Perturbation method, Linear static problem, Linear vibration, Nonlinear static problem, Nonlinear vibration, Stress-strength interference model, Monte Carlo simulation, A new iterative method, Modified iteration formulas, Homotopy perturbation method.

INTRODUCTION

Finite element is a world recognized tool for analyzing structures. The influence of randomness on some structures cannot be ignored. Material properties, geometric parameters and applied loads of the structure have a great impact on dams, buildings, bridges, mechanical parts, *etc*. Considering the influence of random factors, the stochastic finite element is introduced. Stochastic finite element and structural reliability calculation methods are combined to calculate the reliability of the structure.

Two methods are studied for a combination of finite element and reliability methods: the direct method and the quadratic response surface method [1]. An element-free Galerkin method was developed for the reliability analysis of linear-elastic structures [2]. In this paper, finite element reliability methods of the first-order reliability method (FORM) and importance sampling are considered [3]. Analytical methods, combined analytical and simulation-based methods, direct Monte Carlo simulations and the importance sampling strategies are used to analyze dynamic reliability [4]. Five reliability methods calculating the reliability of a

Wenhui Mo

composite structure are given [5]. This paper presents a reliability analysis in geotechnics engineering based on mathematical theories [6]. Reliability of linear structures with parameter uncertainty under non-stationary earthquakes, the perturbation stochastic finite element methodis utilized in deriving reliability of linear structures [7]. Finite element reliability analyses of nonlinear frame structures are employed with sophisticated structural models [8]. In this paper, the reliability of a rotating beam with random properties is studied and a second- order perturbation method is used [9]. Probabilistic risk assessment using finite element analysis for bridge construction is proposed [10]. Advanced Monte Carlo methods for reliability analysis is proposed and uncertainty is regarded as random variables [11]. The context for this paper is FORM in finite element reliability analysis in conjunction with advanced finite element models [12]. The reliability assessment of uncertain linear struc- tures using stochastic finite elements is presented [13]. This paper deals with the reliability analysis using the stochastic finite element method [14]. The objective of this paper is to illustrate an approach for the lifetime reliability assessment of bridges [15]. This paper presents a method for a reliability assessment in structural dynamics [16]. This contribution presents a model reduction technique for reliability sensitivity analysis of nonlinear finite elements [17]. The main aim is to present the stochastic perturbation-based finite element method analysis of the reliability of the underground steel tanks [18]. Improving the reliability of the frequency response function (FRF) by semi-direct model updating is reported [19]. This paper presents a reliability analysis of steady-state seepage by the stochastic scaled boundary finite element method [20]. This paper performs a probabilistic stability analysis for an existing earthfill dam using a stochastic finite element method based on field data [21]. The paper presents a method for reliability analysis of slopes by conditional random finite element method [22]. The objective of the present work is to develop a probabilistic analysis of a Carbon-Nanotube-Reinforced-Polymer (CNRP) material by using the stress-strength model and multiscale finite element model to determine the reliability [23]. The main aim of this paper is to present a reliability estimation procedure for a steel lattice tower based on the stochastic finite element method [24]. This work presents a reliability analysis of structures equipped with friction-based devices [25]. The reliability approach and finite element method are used to estimate failure probability [26]. Dynamic reliability of structure is computed using Successive over Relaxation method or Neumann expansion [30].

In order to improve the calculation accuracy, the calculation formula of third-order perturbation stochastic finite element is presented. The reliability calculation methods of linear static problem, linear vibration, nonlinear static problem and nonlinear vibration are studied using a stochastic finite element method. The stress strength interference model, Monte Carlo simulation, a new iterative method and a modified iteration method by homotopy perturbation method for stochastic finite element reliability calculation are developed. Second order reliability methods for nonlinear static problems and nonlinear vibration are proposed.

Reliability Calculation of Static Problems

The equilibrium equation is written as

$$KU = F \tag{1}$$

where U = the displacement vector, F = the external force, K = the global stiffness matrix.

By applying Taylor series at the mean point $\bar{a} = (\bar{a}_1, \bar{a}_2, \cdots, \bar{a}_n)^T$ of the random variables and the perturbation technology, the following equations are given by

$$U^0 = \left(K^0\right)^{-1} F^0 \tag{2}$$

$$\frac{\partial U}{\partial a_i} = \left(K^0\right)^{-1} \left(\frac{\partial F}{\partial a_i} - \frac{\partial K}{\partial a_i} U^0 \right) \tag{3}$$

$$\frac{\partial^2 U}{\partial a_i \partial a_j} = \left(K^0\right)^{-1} \left(\frac{\partial^2 F}{\partial a_i \partial a_j} - \frac{\partial^2 K}{\partial a_i \partial a_j} U^0 - \frac{\partial K}{\partial a_i} \frac{\partial U}{\partial a_j} - \frac{\partial K}{\partial a_j} \frac{\partial U}{\partial a_i} \right) \tag{4}$$

$$\frac{\partial^3 U}{\partial a_i^2 \partial a_j} = (K^0)^{-1} \left(\frac{\partial^3 F}{\partial a_i^2 \partial a_j} - \frac{\partial^3 K}{\partial a_i^2 \partial a_j} U^0 - 3 \frac{\partial^2 K}{\partial a_i \partial a_j} \frac{\partial U}{\partial a_i} \right.$$

$$\left. -3 \frac{\partial K}{a_i} \frac{\partial^2 U}{\partial a_i \partial a_j} \right) \tag{5}$$

The Taylor expansion formula of U is

$$U = U^0 + \sum_{k=1}^{m} \frac{1}{k!} \sum_{i_1,i_2,\cdots,i_k=1}^{n} \frac{\partial^k U}{\partial a_{i_1} \partial a_{i_2} \cdots \partial a_{i_k}} \left(a^0\right)\left(a_{i_1} - a^0_{i_1}\right)$$

$$\left(a_{i_2} - a^0_{i_2}\right) \cdots \left(a_{i_k} - a^0_{i_k}\right) +$$

$$\frac{1}{(m+1)!} \sum_{i_1,i_2,\cdots,i_{m+1}=1}^{n} \frac{\partial^{m+1} U}{\partial a_{i_1} \partial a_{i_2} \cdots \partial a_{i_k}} \left(a^0\right)\left(a_{i_1} - a^0_{i_1}\right)$$

$$\left(a_{i_2} - a^0_{i_2}\right) \cdots \left(a_{i_k} - a^0_{i_k}\right) \tag{6}$$

The second-order term of the Taylor expansion formula is given by in the literature. In order to improve the calculation accuracy, the third-order Taylor expansion formula is given in this chapter. The third-order Taylor expansion formula of U is

$$U \approx U^0 + \sum_{k=1}^{3} \frac{1}{k!} \sum_{i_1,i_2,\cdots,i_k=1}^{n} \frac{\partial^k U}{\partial a_{i_1} \partial a_{i_2} \cdots \partial a_{i_k}} \left(a^0\right)\left(a_{i_1} - a^0_{i_1}\right)$$

$$\left(a_{i_2} - a^0_{i_2}\right) \cdots \left(a_{i_k} - a^0_{i_k}\right) \tag{7}$$

The mean of U is given by

$$E(U) \approx E(U^0) +$$

$$\sum_{k=1}^{3} \frac{1}{k!} \sum_{i_1,i_2,\cdots,i_k=1}^{n} E\left(\sum_{i_1,i_2,\cdots,i_k=1}^{n} \frac{\partial^k U}{\partial a_{i_1} \partial a_{i_2} \cdots \partial a_{i_k}} \left(a^0\right)\left(a_{i_1} - a^0_{i_1}\right) \right.$$

$$\left. \left(a_{i_2} - a^0_{i_2}\right) \cdots \left(a_{i_k} - a^0_{i_k}\right) \right) \tag{8}$$

where $E(U)$ is to take the expectation operator.

$\operatorname{cov}\left(U_{j_1}, U_{j_2}\right)$ is given by

$$\operatorname{cov}\left(U_{j_1}, U_{j_2}\right) \approx \operatorname{cov}(\sum_{k=1}^{3}\frac{1}{k!}\sum_{i_1,i_2,\cdots,i_k=1}^{n}\frac{\partial^k U_{j_1}}{\partial a_{i_1}\partial a_{i_2}\cdots\partial a_{i_k}}\left(a^0\right)\left(a_{i_1}-a^0_{i_1}\right)$$

$$\left(a_{i_2}-a^0_{i_2}\right)\cdots\left(a_{i_k}-a^0_{i_k}\right),\sum_{k=1}^{3}\frac{1}{k!}\sum_{i_1,i_2,\cdots,i_k=1}^{n}\frac{\partial^k U_{j_2}}{\partial a_{i_1}\partial a_{i_2}\cdots\partial a_{i_k}}\left(a^0\right)\left(a_{i_1}-a^0_{i_1}\right) \qquad (9)$$

$$\left(a_{i_2}-a^0_{i_2}\right)\cdots\left(a_{i_k}-a^0_{i_k}\right))$$

where $\operatorname{cov}\left(U_{j_1}, U_{j_2}\right)$ is the covariance.

$D\left(U_{j_1}\right)$ is given by

$$D\left(U_{j_1}\right) \approx \operatorname{cov}(\sum_{k=1}^{3}\frac{1}{k!}\sum_{i_1,i_2,\cdots,i_k=1}^{n}\frac{\partial^k U_{j_1}}{\partial a_{i_1}\partial a_{i_2}\cdots\partial a_{i_k}}\left(a^0\right)\left(a_{i_1}-a^0_{i_1}\right)$$

$$\left(a_{i_2}-a^0_{i_2}\right)\cdots\left(a_{i_k}-a^0_{i_k}\right),$$

$$\sum_{k=1}^{3}\frac{1}{k!}\sum_{i_1,i_2,\cdots,i_k=1}^{n}\frac{\partial^k U_{j_1}}{\partial a_{i_1}\partial a_{i_2}\cdots\partial a_{i_k}}\left(a^0\right)\left(a_{i_1}-a^0_{i_1}\right) \qquad (10)$$

$$\left(a_{i_2}-a^0_{i_2}\right)\cdots\left(a_{i_k}-a^0_{i_k}\right))$$

where $D\left(U_{j_1}\right)$ is the variance .

The stress for the element d is given by

$$\{\sigma\} = [D][B]U^d$$

$$(11)$$

where, $[D]$ =the material response matrix of element d, $[B]$ =the gradient matrix of the element d and U^d =the element d nodal displacement vector .

The partial derivative of Eq.11 with respect to a_i is given by

$$\frac{\partial \{\sigma\}}{\partial a_i} = \frac{\partial [D]}{\partial a_i}[B]U^d + [D]\frac{\partial [B]}{\partial a_i}U^d + [D][B]\frac{\partial U^d}{\partial a_i}$$

$$(12)$$

The partial derivative of Eq.12 with respect to a_j is given by

$$\frac{\partial^2 \{\sigma\}}{\partial a_i \partial a_j} = \frac{\partial^2 [D]}{\partial a_i \partial a_j}[B]U^d + \frac{\partial [D]}{\partial a_i}\frac{\partial [B]}{\partial a_j}U^d + \frac{\partial [D]}{\partial a_i}[B]\frac{\partial U^d}{\partial a_j}$$

$$+\frac{\partial [D]}{\partial a_j}\frac{\partial [B]}{\partial a_i}U^d + [D]\frac{\partial^2 [B]}{\partial a_i \partial a_j}U^d + [D]\frac{\partial [B]}{\partial a_i}\frac{\partial U^d}{\partial a_j}$$

$$+\frac{\partial [D]}{\partial a_j}[B]\frac{\partial U^d}{\partial a_i} + [D]\frac{\partial [B]}{\partial a_j}\frac{\partial U^d}{\partial a_i} + [D][B]\frac{\partial^2 U^d}{\partial a_i \partial a_j}$$

$$(13)$$

The partial derivative of Eq.13 with respect to a_i is given by

$$\frac{\partial^3 \{\sigma\}}{\partial a_i^2 \partial a_j} = \frac{\partial^3 [D]}{\partial a_i^2 \partial a_j}[B]U^d + \frac{\partial^2 [D]}{\partial a_i \partial a_j}\frac{\partial [B]}{\partial a_i}U^d + \frac{\partial^2 [D]}{\partial a_i \partial a_j}[B]\frac{\partial U^d}{\partial a_i}$$

$$+ \frac{\partial^2 [D]}{\partial a_i^2}\frac{\partial [B]}{\partial a_j}U^d + \frac{\partial [D]}{\partial a_i}\frac{\partial^2 [B]}{\partial a_i \partial a_j}U^d + \frac{\partial [D]}{\partial a_i}\frac{\partial [B]}{\partial a_j}\frac{\partial U^d}{\partial a_i}$$

$$+ \frac{\partial^2 [D]}{\partial a_i^2}[B]\frac{\partial U^d}{\partial a_j} + \frac{\partial [D]}{\partial a_i}\frac{\partial [B]}{\partial a_i}\frac{\partial U^d}{\partial a_j} + \frac{\partial [D]}{\partial a_i}[B]\frac{\partial^2 U^d}{\partial a_i \partial a_j}$$

$$+ \frac{\partial^2 [D]}{\partial a_i \partial a_j}\frac{\partial [B]}{\partial a_i}U^d + \frac{\partial [D]}{\partial a_j}\frac{\partial^2 [B]}{\partial a_i^2}U^d + \frac{\partial [D]}{\partial a_j}\frac{\partial [B]}{\partial a_i}\frac{\partial U^d}{\partial a_i}$$

$$+ \frac{\partial [D]}{\partial a_i}\frac{\partial^2 [B]}{\partial a_i \partial a_j}U^d + [D]\frac{\partial^3 [B]}{\partial a_i^2 \partial a_j}U^d + [D]\frac{\partial^2 [B]}{\partial a_i \partial a_j}\frac{\partial U^d}{\partial a_i}$$

$$+ \frac{\partial [D]}{\partial a_i}\frac{\partial [B]}{\partial a_i}\frac{\partial U^d}{\partial a_j} + [D]\frac{\partial^2 [B]}{\partial a_i^2}\frac{\partial U^d}{\partial a_j} + [D]\frac{\partial [B]}{\partial a_i}\frac{\partial^2 U^d}{\partial a_i \partial a_j}$$

$$+ \frac{\partial^2 [D]}{\partial a_i \partial a_j}[B]\frac{\partial U^d}{\partial a_i} + \frac{\partial [D]}{\partial a_j}\frac{\partial [B]}{\partial a_i}\frac{\partial U^d}{\partial a_i} + \frac{\partial [D]}{\partial a_j}[B]\frac{\partial^2 U^d}{\partial a_i^2}$$

$$+ \frac{\partial [D]}{\partial a_i}\frac{\partial [B]}{\partial a_j}\frac{\partial U^d}{\partial a_i} + [D]\frac{\partial^2 [B]}{\partial a_i \partial a_j}\frac{\partial U^d}{\partial a_i} + [D]\frac{\partial [B]}{\partial a_j}\frac{\partial^2 U^d}{\partial a_i^2}$$

$$+ \frac{\partial [D]}{\partial a_i}[B]\frac{\partial^2 U^d}{\partial a_i \partial a_j} + [D]\frac{\partial [B]}{\partial a_i}\frac{\partial^2 U^d}{\partial a_i \partial a_j} + [D][B]\frac{\partial^3 U^d}{\partial a_i^2 \partial a_j}$$

$$(14)$$

The stress is expanded at the mean value point $\bar{a} = \left(\bar{a}_1, \bar{a}_2, \cdots, \bar{a}_i, \cdots, \bar{a}_{n_1} \right)^T$ by means of a Taylor series. By taking the expectation operator for two sides of the above Eq.11, the mean of stress is obtained as

$$E\left(\sigma \right) \approx E\left(\sigma^0 \right) +$$

$$\sum_{k=1}^{3} \frac{1}{k!} \sum_{i_1, i_2, \cdots, i_k = 1}^{n} E\left(\sum_{i_1, i_2, \cdots, i_k = 1}^{n} \frac{\partial^k \sigma}{\partial a_{i_1} \partial a_{i_2} \cdots \partial a_{i_k}} \left(a^0 \right) \left(a_{i_1} - a^0_{i_1} \right) \right) \tag{15}$$

$$\left(a_{i_2} - a^0_{i_2} \right) \cdots \left(a_{i_k} - a^0_{i_k} \right))$$

where $E\left(\sigma \right)$ is to take the expectation operator.

$\operatorname{cov}\left(\sigma_{j_3}, \sigma_{j_4} \right)$ is given by

$$\operatorname{cov}\left(\sigma_{j_3}, \sigma_{j_4} \right) \approx \operatorname{cov}(\sum_{k=1}^{3} \frac{1}{k!} \sum_{i_1, i_2, \cdots, i_k = 1}^{n} \frac{\partial^k \sigma_{j_3}}{\partial a_{i_1} \partial a_{i_2} \cdots \partial a_{i_k}} \left(a^0 \right) \left(a_{i_1} - a^0_{i_1} \right)$$

$$\left(a_{i_2} - a^0_{i_2} \right) \cdots \left(a_{i_k} - a^0_{i_k} \right),$$

$$\sum_{k=1}^{3} \frac{1}{k!} \sum_{i_1, i_2, \cdots, i_k = 1}^{n} \frac{\partial^k \sigma_{j_4}}{\partial a_{i_1} \partial a_{i_2} \cdots \partial a_{i_k}} \left(a^0 \right) \left(a_{i_1} - a^0_{i_1} \right)$$

$$\left(a_{i_2} - a^0_{i_2} \right) \cdots \left(a_{i_k} - a^0_{i_k} \right)) \tag{16}$$

where $\operatorname{cov}\left(\sigma_{j_3}, \sigma_{j_4} \right)$ is the covariance .

$D\left(\sigma_{j_3} \right)$ is given by

$$D\left(\sigma_{j_3}\right) \approx \mathrm{cov}(\sum_{k=1}^{3}\frac{1}{k!}\sum_{i_1,i_2,\cdots,i_k=1}^{n}\frac{\partial^k \sigma_{j_3}}{\partial a_{i_1}\partial a_{i_2}\cdots\partial a_{i_k}}\left(a^0\right)\left(a_{i_1}-a^0{}_{i_1}\right)$$

$$\left(a_{i_2}-a^0{}_{i_2}\right)\cdots\left(a_{i_k}-a^0{}_{i_k}\right),$$

$$\sum_{k=1}^{3}\frac{1}{k!}\sum_{i_1,i_2,\cdots,i_k=1}^{n}\frac{\partial^k \sigma_{j_3}}{\partial a_{i_1}\partial a_{i_2}\cdots\partial a_{i_k}}\left(a^0\right)\left(a_{i_1}-a^0{}_{i_1}\right)$$

$$\left(a_{i_2}-a^0{}_{i_2}\right)\cdots\left(a_{i_k}-a^0{}_{i_k}\right))$$

(17)

where $D\left(\sigma_{j_3}\right)$ is variance.

It is well known that matrix inversion requires a large number of CPU time. K_0^{-1} is not easy to solve.

Eqs. 2,3,4 and 5 are rewritten as

$$K^0 U^0 = F^0 \tag{18}$$

$$K^0\frac{\partial U}{\partial a_i} = \frac{\partial F}{\partial a_i} - \frac{\partial K}{\partial a_i}U^0 \tag{19}$$

$$K^0\frac{\partial^2 U}{\partial a_i \partial a_j} = \frac{\partial^2 F}{\partial a_i \partial a_j} - \frac{\partial^2 K}{\partial a_i \partial a_j}U^0 - \frac{\partial K}{\partial a_i}\frac{\partial U}{\partial a_j} - \frac{\partial K}{\partial a_j}\frac{\partial U}{\partial a_i} \tag{20}$$

$$K^0\frac{\partial^3 U}{\partial a_i^2 \partial a_j} = \frac{\partial^3 F}{\partial a_i^2 \partial a_j} - \frac{\partial^3 K}{\partial a_i^2 \partial a_j}U^0 - 3\frac{\partial^2 K}{\partial a_i \partial a_j}\frac{\partial U}{\partial a_i}$$

$$-3\frac{\partial K}{\partial a_i}\frac{\partial^2 U}{\partial a_i \partial a_j} \tag{21}$$

Eqs.18, 19, 20 and 21 are linear equations .The Cholesky decomposition is an effective method for solving linear equations. Using the Cholesky decomposition, $U^0, \dfrac{\partial U}{\partial a_i}, \dfrac{\partial^2 U}{\partial a_i \partial a_j}, \dfrac{\partial^3 U}{\partial a_i^2 \partial a_j}$ can be obtained at the mean value point $\overline{a}_e = (\overline{a}_1, \overline{a}_2, \cdots, \overline{a}_e, \cdots, \overline{a}_N)^T$. Using Eqs.8 and 9, the mean and covariance of the displacement can be obtained. Using Eqs. 15 and 16, the mean and variance of the stress can be obtained.

Stress strength interference model is the basic theory of reliability design. The distribution probability density functions of stress and strength are drawn in the same coordinate system, and the two distribution probability density functions partially overlap- interfere, which occurs a lot in engineering. Suppose that stress Y and strength X are normal random variables, and the probability density functions are

$$g(y) = \frac{1}{\sqrt{2\pi}\sigma_y}\exp\left[-\frac{(y-\mu_y)^2}{2\sigma_y^2}\right] \tag{22}$$

$$f(x) = \frac{1}{\sqrt{2\pi}\sigma_x}\exp\left[-\frac{(x-\mu_x)^2}{2\sigma_x^2}\right] \tag{23}$$

where μ_y、μ_x and σ_y、σ_x are the means and standard deviations of Y and X respectively.

Reliability is

$$R = \Phi((\mu_x - \mu_y)/(\sigma_x^2 + \sigma_y^2)^{1/2}) \tag{24}$$

Assuming that the displacement and allowable displacement obey normal distribution, the reliability of stiffness is as follows

$$R_t = \Phi((E([U]) - E(U))/(D([U_j]) + D(U_j))^{1/2}) \tag{25}$$

where $E(U)$ is given by Eq.8, $E([U])$ is mean of allowable displacement, $D(U)$ is given by Eq.10, $D([U])$ is the variance of allowable displacement.

Assuming that the stress and strength obey normal distribution, the reliability of strength is as follows

$$R_s = \Phi((E\left(\left[\sigma_j\right]\right) - E\left(\sigma_j\right)) / (D\left(\left[\sigma_j\right]\right) + D\left(\sigma_j\right))^{1/2})$$

(26)

where $E\left(\left[\sigma_j\right]\right)$ is mean of allowable strength , $E(\sigma)$ is given by Eq.15, $D\left(\left[\sigma_j\right]\right)$ is the variance of strength , $D(\sigma)$ is given by Eq.17.

As long as 12 uniformly distributed random numbers $u_1, u_2, \cdots, u_{12}$ are generated, add them up and subtract 6, the sample value of standard normal variables can be approximately obtained.

And if $X_i \sim N(\mu_i, \sigma_i^2)$, $Z \sim N(0,1)$, using the relation

$$X_i = \mu_i + \sigma_i Z$$

(27)

A general normal random variable can be obtained.

Where $X_i \sim N(\mu_i, \sigma_i^2)$ indicates X_i is submitted to a normal random variable with a mean value of μ_i, and variance of σ_i^2. $Z \sim N(0,1)$ indicates Z is submitted to a normal random variable with a mean value of 0 and variance of 1.If the random variable follows other distributions, the sample value of the random variable can be obtained by inverse transformation.

Material properties, geometric parameters and loads are regarded as random variables. They are expressed as $a_1, a_2, \cdots, a_i, \cdots, a_{n_1}$.

When they obey arbitrary distribution, the samples are generated by the following method.

$$P\left\{\left|x-\mu'\right|\geq\varepsilon\right\}\leq\frac{\sigma^2}{\varepsilon^2} \tag{28}$$

where μ' is the mean, σ is the standard deviation, ε is any positive number, and the above formula is called Chebyshev inequality.

Chebyshev inequality can be rewritten as

$$P\left\{\left|x-\mu'\right|<\varepsilon\right\}\geq1-\frac{\sigma^2}{\varepsilon^2} \tag{29}$$

To take $\varepsilon=10\sigma_i$, $x=a_i$,   we get

$$P\left\{\left|a_i-\mu_i\right|<10\sigma_i\right\}\geq0.99 \tag{30}$$

then

$$a_i<10\sigma_i+\mu_i \tag{31}$$

and

$$a_i>-10\sigma_i+\mu_i \tag{32}$$

The application of Monte Carlo simulation in reliability calculation is to randomly select a group of values from the random variable distribution, substitute them into the stress calculation formula to obtain a stress value, and then compare it with a strength value extracted from the strength distribution. If the stress is greater than the strength, it will fail; On the contrary, it is safe. If the number of simulations is *N*, the number of failures is *F*.

Failure probability is

$$P_f=\frac{F}{N} \tag{33}$$

The reliability is

$$R_N = 1 - \frac{F}{N} \tag{34}$$

Given the mean and variance of the allowable displacement, the mean and variance of the displacement are calculated by Eqs.8,10. The mean and variance of allowable stress are known, and the mean and variance of stress are calculated by Eqs.15,17. The reliability of stiffness and strength can be obtained by using the above Monte Carlo simulation method.

Eq.1 is rewritten as

Ax=b, $\tag{35}$

In this section, a new iterative method (NIM)and its convergence analysis for finding a solution of Eq.35 ,along with the estimation of error bounds ,are described. Let $A \in Cm \times n$ and T be a subspace of Cn. Starting with Z0=βY, where β is a non zero realscalar,$Y \in Cn \times m$ satisfying R(Y) $\subseteq$ T and for any x0 $\in$ _T,the iterative method [27] is defined for k=0,1,2,... by

Zk+1=Zk(2I−AZk) $\tag{36}$

xk+1=xk+Zk+1(b−Axk) $\tag{37}$

Vectors $U_1, U_2, \cdots, U_{N_1}$ are solutions of N_1 Eqs.35. Representing $U_1, U_2, \cdots, U_{N_1}$ and samples of stochastic variables into Eq.11, $\sigma_1, \sigma_2, \cdots, \sigma_{N_1}$ are obtained. The stiffness and strength reliability can be obtained using the Monte Carlo simulation method.

Structural Reliability Calculation of Linear Vibration

Material properties, geometry parameters and applied loads of structures are considered normal random variables, and are indicated as $a_1, a_2, \cdots, a_i, \cdots, a_n$. Their means are $\mu_1, \mu_2, \cdots, \mu_i, \cdots, \mu_n$, and their variances are $\sigma_1^2, \sigma_2^2, \cdots, \sigma_i^2, \cdots, \sigma_n^2$.

For a linear system, the dynamic equilibrium equation is given by

$$[M]\{\ddot{U}\} + [C]\{\dot{U}\} + [K]\{U\} = \{F\} \tag{38}$$

where $\{\ddot{U}\},\{\dot{U}\},\{U\}$ are the acceleration, velocity and displacement vectors. $[M],[K]$ and $[C]$ are the global mass, stiffness and damping matrices obtained by assembling the element variables in the global coordinate system.

By using the Newmark method, Eq.38 becomes

$$\{U_{t+\Delta t}\} = \left[\tilde{K}\right]^{-1}\{\tilde{F}_{t+\Delta t}\} \tag{39}$$

where, $\{U_{t+\Delta t}\}$, $\left[\tilde{K}\right]$ and $\{\tilde{F}_{t+\Delta t}\}$ indicate the displacement vector,

stiffness matrix and load vector at time $t+\Delta t$.

Eq.39 can be rewritten

$$\left[\tilde{K}_{t+\Delta t}\right]\{U_{t+\Delta t}\} = \{\tilde{F}_{t+\Delta t}\} \tag{40}$$

The partial derivative of Eq.40 with respect to a_i is given by

$$\frac{\partial\{U_{t+\Delta t}\}}{\partial a_i} = \left[\tilde{K}\right]^{-1}\left(\frac{\partial\{\tilde{F}_{t+\Delta t}\}}{\partial a_i} - \frac{\partial\left[\tilde{K}\right]}{\partial a_i}\{U_{t+\Delta t}\}\right) \tag{41}$$

where

$$\frac{\partial\{\tilde{F}_{t+\Delta t}\}}{\partial a_i} = \frac{\partial\{F_{t+\Delta t}\}}{\partial a_i} + \frac{\partial[M]}{\partial a_i}(b_0\{U_t\}$$

$$+b_2\{\dot{U}_t\}+b_3\{\ddot{U}_t\})+$$

$$[M]\left(b_0\frac{\partial\{U_t\}}{\partial a_i}+\right.$$

$$b_2 \frac{\partial \{\dot{U}_t\}}{\partial a_i} + b_3 \frac{\partial \{\ddot{U}_t\}}{\partial a_i} \Bigg)$$

$$+ \frac{\partial [C]}{\partial a_i} (b_1 \{U_t\} + b_4 \{\dot{U}_t\} + b_5 \{\ddot{U}_t\})$$

$$+ [C] \left(b_1 \frac{\partial \{U_t\}}{\partial a_i} + b_4 \frac{\partial \{\dot{U}_t\}}{\partial a_i} + b_5 \frac{\partial \{\ddot{U}_t\}}{\partial a_i} \right)$$

(42)

After $\dfrac{\partial \{U_t\}}{\partial a_i} = q_0, \dfrac{\partial \{\dot{U}_t\}}{\partial a_i} = \dot{q}_0, \dfrac{\partial \{\ddot{U}_t\}}{\partial a_i} = \ddot{q}_0$ are given, Eq.41 can be calculated.

The partial derivative of Eq.41 with respect to a_j is given by

$$\frac{\partial^2 \{U_{t+\Delta t}\}}{\partial a_i \partial a_j} = [\tilde{K}]^{-1} \left(\frac{\partial^2 \{\tilde{F}_{t+\Delta t}\}}{\partial a_i \partial a_j} - \frac{\partial [\tilde{K}]}{\partial a_i} \frac{\partial \{U_{t+\Delta t}\}}{\partial a_j} \right.$$

$$\left. - \frac{\partial [\tilde{K}]}{\partial a_j} \frac{\partial \{U_{t+\Delta t}\}}{\partial a_i} - \frac{\partial^2 [\tilde{K}]}{\partial a_i \partial a_j} \{U_{t+\Delta t}\} \right)$$

(43)

where

$$\frac{\partial^2 \{\tilde{F}_{t+\Delta t}\}}{\partial a_i \partial a_j} = \frac{\partial^2 \{F_{t+\Delta t}\}}{\partial a_i \partial a_j} +$$

$$\frac{\partial^2 [M]}{\partial a_i \partial a_j} (b_0 \{U_t\} + b_2 \{\dot{U}_t\} + b_3 \{\ddot{U}_t\})$$

$$+\frac{\partial[M]}{\partial a_i}(b_0\frac{\partial\{U_t\}}{\partial a_j}+b_2\frac{\partial\{\dot{U}_t\}}{\partial a_j}+b_3\frac{\partial\{\ddot{U}_t\}}{\partial a_j})$$

$$+\frac{\partial[M]}{\partial a_j}(b_0\frac{\partial\{U_t\}}{\partial a_i}+b_2\frac{\partial\{\dot{U}_t\}}{\partial a_i}+b_3\frac{\partial\{\ddot{U}_t\}}{\partial a_i})$$

$$+[M]\left(b_0\frac{\partial^2\{U_t\}}{\partial a_i\partial a_j}+b_2\frac{\partial^2\{\dot{U}_t\}}{\partial a_i\partial a_j}+b_3\frac{\partial^2\{\ddot{U}_t\}}{\partial a_i\partial a_j}\right)$$

$$+\frac{\partial^2[C]}{\partial a_i\partial a_j}(b_1\{U_t\}+b_4\{\dot{U}_t\}+b_5\{\ddot{U}_t\})$$

$$+\frac{\partial[C]}{\partial a_i}(b_1\frac{\partial\{U_t\}}{\partial a_j}+b_4\frac{\partial\{\dot{U}_t\}}{\partial a_j}+b_5\frac{\partial\{\ddot{U}_t\}}{\partial a_j})$$

$$+\frac{\partial[C]}{\partial a_j}(b_1\frac{\partial\{U_t\}}{\partial a_i}+b_4\frac{\partial\{\delta\dot{U}_t\}}{\partial a_i}+b_5\frac{\partial\{\ddot{U}_t\}}{\partial a_i})$$

$$+[C]\left(b_1\frac{\partial^2\{U_t\}}{\partial a_i\partial a_j}+b_4\frac{\partial^2\{\dot{U}_t\}}{\partial a_i\partial a_j}+b_5\frac{\partial^2\{\ddot{U}_t\}}{\partial a_i\partial a_j}\right)$$

(44)

After $\dfrac{\partial\{U_t\}}{\partial a_j}=q_1$, $\dfrac{\partial\{\dot{U}_t\}}{\partial a_j}=\dot{q}_1$, $\dfrac{\partial\{\ddot{U}_t\}}{\partial a_j}=\ddot{q}_1$, $\dfrac{\partial^2\{U_t\}}{\partial a_i\partial a_j}=r_0$, $\dfrac{\partial^2\{\dot{U}_t\}}{\partial a_i\partial a_j}=\dot{r}_0$, $\dfrac{\partial^2\{\ddot{U}_t\}}{\partial a_i\partial a_j}=\ddot{r}_0$

are given, Eq.43 can be calculated.

The displacement is expanded at the mean value point $\bar{a}=\left(\bar{a}_1,\bar{a}_2,\cdots,\bar{a}_i,\cdots,\bar{a}_{n_1}\right)^T$ by means of a Taylor series. The mean of $\delta_{t+\Delta t}$ is obtained as

$$\mu\{U_{t+\Delta t}\} \approx \{U_{t+\Delta t}\}\big|_{a=\bar{a}} + \frac{1}{2}\sum_{i=1}^{N}\sum_{j=1}^{N}\frac{\partial^2\{U_{t+\Delta t}\}}{\partial a_i \partial a_j}\big|_{a=\bar{a}} Cov(a_i, a_j) \tag{45}$$

where, $\mu\{U_{t+\Delta t}\}$ expresses mean value $\delta_{t+\Delta t}$ and $Cov(a_i, a_j)$ is the

covariance between a_i and a_j.

The variance of $\delta_{t+\Delta t}$ is given by

$$Var\{U_{t+\Delta t}\} \approx \sum_{i=1}^{N}\sum_{j=1}^{N}\frac{\partial\{U_{t+\Delta t}\}}{\partial a_i}\big|_{a=\bar{a}} .$$

$$\frac{\partial\{U_{t+\Delta t}\}}{\partial a_j}\big|_{a=\bar{a}} \cdot Cov(a_i, a_j) \tag{46}$$

The partial derivative of $\ddot{U}_{t+\Delta t}$ with respect to a_i is given by

$$\frac{\partial\{\ddot{U}_{t+\Delta t}\}}{\partial a_i} = b_0\left(\frac{\partial\{U_{t+\Delta t}\}}{\partial a_i} - \frac{\partial\{U_t\}}{\partial a_i}\right)$$

$$-b_2\frac{\partial\{\dot{U}_t\}}{\partial a_i} - b_3\frac{\partial\{\ddot{U}_t\}}{\partial a_i} \tag{47}$$

The partial derivative of $\dot{U}_{t+\Delta t}$ with respect to a_i is given by

$$\frac{\partial\{\dot{U}_{t+\Delta t}\}}{\partial a_i} = \frac{\partial\{\dot{U}_t\}}{\partial a_i} + b_6\frac{\partial\{\ddot{U}_t\}}{\partial a_i} + b_7\frac{\partial\{\ddot{U}_{t+\Delta t}\}}{\partial a_i} \tag{48}$$

The partial derivative of Eq.47 with respect to a_j is given by

$$\frac{\partial^2 \{\ddot{U}_{t+\Delta t}\}}{\partial a_i \partial a_j} = b_0 \left(\frac{\partial^2 \{U_{t+\Delta t}\}}{\partial a_i \partial a_j} - \frac{\partial^2 \{U_t\}}{\partial a_i \partial a_j} \right) -$$

$$b_2 \frac{\partial^2 \{\dot{U}_t\}}{\partial a_i \partial a_j} - b_3 \frac{\partial^2 \{\ddot{U}_t\}}{\partial a_i \partial a_j} \tag{49}$$

The partial derivative of Eq.48 with respect to a_j is given by

$$\frac{\partial^2 \{\dot{U}_{t+\Delta t}\}}{\partial a_i \partial a_j} = \frac{\partial^2 \{\dot{U}_t\}}{\partial a_i \partial a_j} + b_6 \frac{\partial^2 \{\ddot{U}_t\}}{\partial a_i \partial a_j} + b_7 \frac{\partial^2 \{\ddot{U}_{t+\Delta t}\}}{\partial a_i \partial a_j} \tag{50}$$

Eqs 47, 48 , 49 and 50 must be calculated for the following iteration.

Then, the mean and variance of displacement are obtained at time $t + i_1 \Delta t \left(i_1 = 2, 3, \cdots, n_1 \right)$ step by step.

At time $t' = t + i_2 \Delta t \left(i_2 = 1, 2, \cdots, n_1 \right)$, the stress for the element d is given by

$$\{\sigma\} = [D][B]\{U_{t'}{}^d\} \tag{51}$$

where, $[D]$ =the material response matrix of element d , $[B]$=the gradient matrix of the element d and $\{U_{t'}{}^d\}$ =the element d nodal displacement vector at a time t' .

The partial derivative of Eq.51 with respect to a_i is given by

$$\frac{\partial \{\sigma\}}{\partial a_i} = \frac{\partial [D]}{\partial a_i}[B]\{U_{t'}{}^d\} + [D]\frac{\partial [B]}{\partial a_i}\{U_{t'}{}^d\}$$

$$+[D][B]\frac{\partial\{U_{t'}^{d}\}}{\partial a_i} \tag{52}$$

The partial derivative of Eq.52 with respect to a_j is given by

$$\frac{\partial^2\{\sigma\}}{\partial a_i\partial a_j} = \frac{\partial^2[D]}{\partial a_i\partial a_j}[B]\{U_{t'}^{d}\} +$$

$$\frac{\partial[D]}{\partial a_i}\frac{\partial[B]}{\partial a_j}\{U_{t'}^{d}\} + \frac{\partial[D]}{\partial a_i}[B]\frac{\partial\{U_{t'}^{d}\}}{\partial a_j}$$

$$+\frac{\partial[D]}{\partial a_j}\frac{\partial[B]}{\partial a_i}\{U_{t'}^{d}\} + [D]\frac{\partial^2[B]}{\partial a_i\partial a_j}\{U_{t'}^{d}\}$$

$$+[D]\frac{\partial[B]}{\partial a_i}\frac{\partial\{U_{t'}^{d}\}}{\partial a_j}$$

$$+\frac{\partial[D]}{\partial a_j}[B]\frac{\partial\{U_{t'}^{d}\}}{\partial a_i} +$$

$$[D]\frac{\partial[B]}{\partial a_j}\frac{\partial\{U_{t'}^{d}\}}{\partial a_i} + [D][B]\frac{\partial^2\{U_{t'}^{d}\}}{\partial a_i\partial a_j} \tag{53}$$

The stress is expanded at the mean value point $\bar{a} = \left(\bar{a}_1, \bar{a}_2, \cdots, \bar{a}_i, \cdots, \bar{a}_{n_1}\right)^T$ by means of a Taylor series. By taking the expectation operator for two sides of the above Eq.51, the mean of stress is obtained as

$$\mu\{\sigma\} \approx \{\sigma\}\big|_{a=\bar{a}} +$$

$$\frac{1}{2}\sum_{i=1}^{N}\sum_{j=1}^{N}\frac{\partial^2\{\sigma\}}{\partial a_i\partial a_j}\bigg|_{a=\bar{a}}Cov(a_i, a_j) \tag{54}$$

where, $\mu\{\sigma\}$ expresses the mean of σ and $Cov(a_i, a_j)$ is the covariance between a_i and a_j.

The variance of σ is given by

$$Var\{\sigma\} \approx \sum_{i=1}^{N}\sum_{j=1}^{N}\frac{\partial\{\sigma\}}{\partial a_i}\Big|_{a=\bar{a}} \cdot$$

$$\frac{\partial\{\sigma\}}{\partial a_j}\Big|_{a=\bar{a}}\cdot Cov(a_i, a_j) \tag{55}$$

Using the stress strength interference model, the reliability of stiffnes is

$$R_u = \Phi(([U]_{t'}\} - \{U_{t'}\})/(\mathrm{var}\{[U]_{t'}\} + \mathrm{var}\{U_{t'}\})^{1/2}) \tag{56}$$

where $\{[U]_{t'}\}$ is the mean of allowable displacement, $\{U_{t'}\}$ is the mean of displacement, $\mathrm{var}\{[U]_{t'}\}$ is the variance of allowable displacement and $\mathrm{var}\{U_{t'}\}$ is the variance of displacement. Assuming that the stress and strength obey normal distribution, the reliability of strength is as follows

$$R_{ss} = \Phi((\mu([\sigma_j]) - \mu(\sigma_j))/(\mathrm{var}([\sigma_j]) + \mathrm{var}(\sigma_j))^{1/2}) \tag{57}$$

where $\mu([\sigma_j])$ is the mean of strength, $\mu(\sigma_j)$ is the mean of stress, $\mathrm{var}([\sigma_j])$ is the variance of strength, $\mathrm{var}(\sigma_j)$ is the variance ofstress.

The mean and variance of the allowable displacement are given. The mean and variance of the displacement are calculated by Eqs.45, 46. The mean and variance of allowable stress are known, and the mean and variance of stress are calculated by Eqs.54, 55. The reliability of stiffness and strength can be obtained by using the above Monte Carlo simulation or a new iterative method (NIM) [27].

Reliability of Nonlinear Structures

N_1 samples of the vector $\bar{a}$ are produced. N_1 matrices $\left[\tilde{K}\right]$ and N_1 Eq.1 are generated. For nonlinear structure, Eq.1 is a system of nonlinear equations. A modified iteration formulas by the homotopy perturbation method (MIHPD)with accelerated fourth-and fifth-order convergence [28]is used to solve Eq.1 .Eq.1 is rewritten as

$$\Phi(X)=\begin{cases} f_1(X) \\ f_2(X) \\ \vdots \\ f_N(X) \\ X=(X_1,X_2,\cdots X_N)^T \in \Re^N \end{cases} \tag{58}$$

The solution of Eq.58 is given by $y_{i,m} = x_{i,m} - \sum_{n=1}^{N} \Psi_{i,n}(X_m)f_n(X_m)$ (59)

$$z_{i,m} = -\sum_{n=1}^{N} \Psi_{i,n}(Y_m)f_n(Y_m) \tag{60}$$

$$x_{i,m+1} = y_{i,m} + z_{i,m} - \frac{1}{2}\sum_{n=1}^{N}\sum_{j=1}^{N}\sum_{k=1}^{N} \Psi_{i,n}(X_m)\frac{\partial^2 f_n(Y_m)}{\partial x_j \partial x_k} z_{j,m} z_{k,m} \tag{61}$$

$i = 1,2,\cdots,N, m=0,1,\cdots$

Vectors $U_1,U_2,\cdots,U_{N_1}$ are solutions of N_1 Eqs.1 The reliability of stiffness is also calculated by the Monte Carlo simulation.

The mean values of elastic modulus, Poisson's ratio, geometry parameters and load are substituted into Eq.1, and the mean values of displacement and stress are obtained by using the incremental tangent stiffness method and the initial stress method. The mean value of stress is substituted into the following equation (Eq.62), which contains only random variables, such as elastic modulus, Poisson's ratio and geometry parameters. The mean value and covariance of stress are calculated as

$$\{\sigma\} = [D]_{ep}[B]U$$

(62)

where $[D]_{ep}$ is elastoplastic matrix.

The stress for the element d is given by

(63)

$$\{\sigma\} = [D]_{ep}[B]U^d$$

where $[B]$=the gradient matrix of element d and U^d =the element d nodal displacement vector.

Substituting N_1 samples of vectors $\bar{a}$ and N_1 U^d into Eq.63,

vectors $\{\sigma\}_1, \{\sigma\}_2, \cdots, \{\sigma\}_{N_1}$ can be obtained.

The mean of $\{\sigma\}$ is given by

$$\mu\{\sigma\} = \frac{\{\sigma\}_1 + \{\sigma\}_2 + \cdots + \{\sigma\}_{N_1}}{N_1}$$

(64)

The variance of $\{\sigma\}$ is given by

$$Var\{\sigma\} = \frac{1}{N_1 - 1}\sum_{i=1}^{N_1}\left(\{\sigma\}_i - \mu\{\sigma\}\right)^2$$

(65)

The reliability of strength can be calculated by the Monte Carlo simulation.

The reliability of strength can be calculated by the Second order reliability method.

Stress is given by

$$\begin{pmatrix} \sigma_x \\ \sigma_y \\ \sigma_z \\ \tau_{xy} \\ \tau_{yz} \\ \tau_{zx} \end{pmatrix}$$

The stress for the element d is given by

$$\{\sigma\} = [D]_{ep}[B]U^d \tag{66}$$

Allowable stress is given by

$$\begin{pmatrix} [\sigma_x] \\ [\sigma_y] \\ [\sigma_z] \\ [\tau_{xy}] \\ [\tau_{yz}] \\ [\tau_{zx}] \end{pmatrix}$$

Take σ_x as an example to calculate the reliability, and the rest are the same.

The limit state function is given by

$$g(x<0) = \sigma_x - [\sigma_x] \le 0 \tag{67}$$

The probability density function is given by

$$f(x_1, x_2, \cdots, x_n) = \frac{1}{(2\pi)^{n/2}} \exp\left\{ \sum_{i=1}^{n} x_i^2 \right\} \tag{68}$$

The probability of failure is [29]

$$P\left(g\left(x<0\right)\right) \approx \Phi\left(-\beta_0\right)\left|\sum_{i=1}^{k}\left(\prod_{j=1}^{n-1}\left(1-\beta_0\overline{\kappa}_{i,j}\right)^{-\frac{1}{2}}\right)\right|$$

(69)

where, $\overline{\kappa}_{i,j}$ is the ordered curature of the surface $G = [x; g(x) = 0]$ at $\underline{z}_i$.

Reliability of strength is defined as

$$R_{S1} = 1 - P\left(g\left(x<0\right)\right)$$

(70)

Reliability calculation of nonlinear vibration

By using the Newmark method, Eq.39 can be rewritten

$$\left[\tilde{K}\right]\left\{U_{t+\Delta t}\right\} = \left\{\tilde{F}_{t+\Delta t}\right\}$$

(71)

The stiffness matrix is defined as

$$\left[\tilde{K}\right] = \left[K\right] + b_0\left[M\right] + b_1\left[C\right]$$

(72)

At the time $t + \Delta t$, the load vector is defined as

$$\left\{\tilde{F}_{t+\Delta t}\right\} = \left\{F_{t+\Delta t}\right\} + \left[M\right]\left(b_0\left\{U_t\right\}+\right.$$

$$\left. b_2\left\{\dot{U}_t\right\} + b_3\left\{\ddot{U}_t\right\}\right) + \left[C\right]\left(b_1\left\{U_t\right\} + b_4\left\{\dot{U}_t\right\} + b_5\left\{\ddot{U}_t\right\}\right)$$

(73)

At the time $t + \Delta t$, the acceleration vector and velocity vector are obtained as

$$\left\{\ddot{U}_{t+\Delta t}\right\} = b_0\left(\left\{U_{t-\Delta t}\right\} - \left\{U_t\right\}\right) - b_2\left\{\dot{U}_t\right\} - b_3\left\{\ddot{U}_t\right\}$$

(74)

$$\left\{\dot{U}_{t+\Delta t}\right\} = \left\{\dot{U}_t\right\} + b_6\left\{\ddot{U}_t\right\} + b_7\left\{\ddot{U}_{t-\Delta t}\right\}$$

(75)

Vectors $\{U_{t+i_1\Delta t}\}, \{\dot{U}_{t+i_1\Delta t}\}, \{\ddot{U}_{t+i_1\Delta t}\}$ are solved at time $t+i_1\Delta t\,(i_1=2,3,\cdots,n_1)$ step-by -step.

For nonlinear vibration, Eq.71 is a system of nonlinear equations. N_1 samples of vector $\bar{a}$ are produced. N_1 matrices $\left[\tilde{K}_{t+i_1\Delta t}\right]$ and N_1 Eq.71 are generated. A modified iteration formulas by the homotopy perturbation method (MIHPD) with accelerated fourth-and fifth-order convergence [28] is used to solve Eq.71. Vectors $\{U_{t+\Delta t}\}_1, \{U_{t+\Delta t}\}_2, \cdots, \{U_{t+\Delta t}\}_{N_1}$ are solutions of N_1 Eqs.71 .

Eqs 72, 73 , 74 and 75 must be calculated for the following iteration.

Then, displacement is obtained at a time $t+i_1\Delta t\,(i_1=2,3,\cdots,n_1)$ step by step. The reliability of stiffness is also calculated by the Monte Carlo simulation.

The mean value of stress is substituted into the following equation (Eq.76), which contains only random variables, such as elastic modulus, Poisson's ratio and geometry parameters.

At the time $t'=t+i_2\Delta t\,(i_2=1,2,\cdots,n_1)$, the stress for the element d is given by

$$\{\sigma\}=[D]_{ep}[B]\{U_{t'}^d\}$$

(76)

Substituting N_1 samples of vectors $\bar{a}$ and N_1 $\{U_{t'}^d\}$ into Eq.76, vectors $\{\sigma\}_1, \{\sigma\}_2, \cdots, \{\sigma\}_{N_1}$ can be obtained.

The mean of $\{\sigma\}$ is given by

$$\mu\{\sigma\}=\frac{\{\sigma\}_1+\{\sigma\}_2+\cdots+\{\sigma\}_{N_1}}{N_1}$$

(77)

The variance of $\{\sigma\}$ is given by

$$Var\{\sigma\} = \frac{1}{N_1 - 1} \sum_{i=1}^{N_1} \left(\{\sigma\}_i - \mu\{\sigma\} \right)^2 \tag{78}$$

The reliability of strength can be calculated by the Monte Carlo simulation.

The reliability of strength can be calculated by the Second order reliability method. See above section.

CONCLUDING REMARKS

The calculation formula of third-order perturbation stochastic finite element is derived in detail. Using the stress-strength interference model and Monte Carlo simulation of structural reliability calculation, the reliability calculation methods of the stochastic finite element are studied. The reliability calculation methods of linear static problem, linear vibration, nonlinear static problem and nonlinear vibration are proposed. The reliability of strength for nonlinear static problems and nonlinear vibration can be calculated by the Second order reliability method. The methods in this chapter can be used to calculate the reliability of dams, buildings, bridges, mechanical parts, *etc*.

REFERENCES

[1]   M. Pendola, A. Mohamed, M. Lemaire, and P. Hornet, "Combination of finite element and reliability methods in nonlinear fracture mechanics", *Reliab. Eng. Syst. Saf.,* vol. 70, pp. 15-27, 2000.

http://dx.doi.org/10.1016/S0951-8320(00)00043-0

[2]   S. Rahman, and B.N. Rao, "An element-free Galerkin method for probabilistic mechanics and reliability", *Int. J. Solids Struct.,* vol. 38, pp. 9313-9330, 2001.

http://dx.doi.org/10.1016/S0020-7683(01)00193-7

[3]   ArmenDer Kiureghian, "Comparison of finite element reliability methods", *Probab. Eng. Mech.,* vol. 17, pp. 337-348, 2002.

http://dx.doi.org/10.1016/S0266-8920(02)00031-0

[4]   S.Manohar, "Dynamic Stiffness Method For Circular Stochastic Timoshenko Beams: Response Variability And Reliability Analyses", *J. Sound Vibrat.,* vol. 253, pp. 1051-1085, 2002.

http://dx.doi.org/10.1006/jsvi.2001.4082

[5]   M. Di Sciuva, and D. Lomario, "A comparison between Monte Carlo and FORMs in calculating the reliability of a composite structure", *Compos. Struct.,* vol. 59, pp. 155-162, 2003.

http://dx.doi.org/10.1016/S0263-8223(02)00170-8

[6]   H.F. Schweiger, and G.M. Peschl, "Reliability analysis in geotechnics with the random set finite element method", *Comput. Geotech.,* vol. 32, pp. 422-435, 2005.

http://dx.doi.org/10.1016/j.compgeo.2005.07.002

[7]   A. Chaudhuri, and S. Chakraborty, "Reliability of linear structures with parameter uncertainty under non-stationary earthquake", *Struct. Saf.,* vol. 28, 2006.

http://dx.doi.org/10.1016/j.strusafe.2005.07.001

[8]   T. Haukaas, and M.H. Scott, "Shape sensitivities in the reliability analysis of nonlinear frame structures", *Comput. Struc.,* vol. 84, no. 15–16, pp. 964-977, 2006.

http://dx.doi.org/10.1016/j.compstruc.2006.02.014

[9]   S.A.A. Hosseini, and S.E. Khadem, "Vibration and reliability of a rotating beam with random properties under random excitation", *Int. J. Mech. Sci.,* vol. 49, pp. 1377-1388, 2007.

http://dx.doi.org/10.1016/j.ijmecsci.2007.04.008

[10] T. Cho, and T.S. Kim, "Probabilistic risk assessment for the construction phases of a bridge construction based on finite element analysis", *Finite Elem. Anal. Des.,* vol. 44, pp. 383-400, 2008.

http://dx.doi.org/10.1016/j.finel.2007.12.004

[11] M.F. Pellissetti, and G.I. Schuëller, "Scalable uncertainty and reliability analysis by integration of advanced Monte Carlo simulation and generic finite element solvers", *Comput. Struc.,* vol. 87, pp. 930-947, 2009.

http://dx.doi.org/10.1016/j.compstruc.2009.04.003

[12] S.D. Koduru, and T. Haukaas, "Feasibility of FORM in finite element reliability analysis", *Struct. Saf.,* vol. 32, pp. 145-153, 2010.

http://dx.doi.org/10.1016/j.strusafe.2009.10.001

[13] H.J. Pradlwarter, and G.I. Schuëller, "Uncertain linear structural systems in dynamics: Efficient stochastic reliability assessment", *Comput. Struc.,* 2010.

http://dx.doi.org/10.1016/j.compstruc.2009.06.010

[14] K. Farah, M. Ltifi, and H. Hassis, "Reliability analysis of slope Stability using stochastic finite element method", *Procedia Eng.,* vol. 10, pp. 1402-1407, 2011.

http://dx.doi.org/10.1016/j.proeng.2011.04.233

[15] N.M. Okasha, D.M. Frangopol, and A.D. Orcesi, "Automated finite element updating using strain data for the lifetime reliability assessment of bridges", *Reliab. Eng. Syst. Saf.,* vol. 99, pp. 139-150, 2012.

http://dx.doi.org/10.1016/j.ress.2011.11.007

[16] B. Goller, H.J. Pradlwarter, and G.I. Schuëller, "Reliability assessment in structural dynamics", *J. Sound Vibrat.,* vol. 332, pp. 2488-2499, 2013.

http://dx.doi.org/10.1016/j.jsv.2012.11.021

[17] H.A. Jensen, F. Mayorga, and C. Papadimitriou, "Reliability sensitivity analysis of stochastic finite element models", *Comput. Methods Appl. Mech. Eng.,* vol. 296, pp. 327-351, 2015.

http://dx.doi.org/10.1016/j.cma.2015.08.007

[18] M. Kamiński, and P. Świta, "Structural stability and reliability of the underground steel tanks with the Stochastic Finite Element Method", *Arch. Civ. Mech. Eng.,* vol. 15, pp. 593-602, 2015.

http://dx.doi.org/10.1016/j.acme.2014.04.010

[19] J.H. Lim, D-S. Hwang, D. Sohn, and J-G. Kim, "Improving the reliability of the frequency response function through semi-direct finite element model updating", *Aerosp. Sci. Technol.,* vol. 54, pp. 59-71, 2016.

http://dx.doi.org/10.1016/j.ast.2016.04.003

[20] A. Johari, and A. Heydari, "Reliability analysis of seepage using an applicable procedure based on stochastic scaled boundary finite element method", *Eng. Anal. Bound. Elem.,* vol. 94, pp. 44-59, 2018.

http://dx.doi.org/10.1016/j.enganabound.2018.05.015

[21] A. Mouyeaux, C. Carvajal, P. Bressolette, L. Peyras, P. Breul, and C. Bacconnet, "Probabilistic stability analysis of an earth dam by Stochastic Finite Element Method based on field data", *Comput. Geotech.,* vol. 101, pp. 34-47, 2018.

http://dx.doi.org/10.1016/j.compgeo.2018.04.017

[22] A. Johari, and A. Gholampour, "A practical approach for reliability analysis of unsaturated slope by conditional random finite element method", *Comput. Geotech.,* vol. 102, pp. 79-91, 2018.

http://dx.doi.org/10.1016/j.compgeo.2018.06.004

[23] J.A. Palacios, and R. Ganesan, "Reliability evaluation of Carbon-Nanotube-Reinforced-Polymer composites based on multiscale finite element model", *Compos. Struct.,* vol. 229, 2019.111381

http://dx.doi.org/10.1016/j.compstruct.2019.111381

[24] J. Szafran, K. Juszczyk, and M. Kamiński, "Experiment-based reliability analysis of structural joints in a steel lattice tower", *J. Construct. Steel Res.,* vol. 154, pp. 278-292, 2019.

http://dx.doi.org/10.1016/j.jcsr.2018.11.006

[25] H.A. Jensen, F. Mayorga, and M.A. Valdebenito, "On the reliability of structures equipped with a class of friction-based devices under stochastic excitation", *Comput. Methods Appl. Mech. Eng.,* vol. 364, 2020.112965

http://dx.doi.org/10.1016/j.cma.2020.112965

[26] M. Taherinasab, and A. Ali, "Aghakouchak, "Estimating failure probability of IBBC connection using direct coupling of reliability approach and finite element method", *J. Build. Eng.,* vol. 38, 2021.102207

http://dx.doi.org/10.1016/j.jobe.2021.102207

[27] S. Srivastava, and D.K. Gupta, "An iterative method for solving general restricted linear equations", *Appl. Math. Comput.,* vol. 262, pp. 344-353, 2015.

http://dx.doi.org/10.1016/j.amc.2015.04.047

[28] K. Sayevand, and H. Jafari, "On systems of nonlinear equations: some modified iteration formulas by the homotopy perturbation method with accelerated fourth-and fifth-order convergence", *Appl. Math. Model.,* vol. 40, pp. 1467-1476, 2016.

http://dx.doi.org/10.1016/j.apm.2015.06.030

[29] K. Breitung, "Asymptotic approximations for multinormal integrals", *J. Eng. Mech.,* vol. 110, pp. 357-366, 1984.

http://dx.doi.org/10.1061/(ASCE)0733-9399(1984)110:3(357)

[30] W. Mo, *Reliability calculations with the stochastic finite element.,* Bentham Science Publishers: Singapore, 2020.

http://dx.doi.org/10.2174/97898114855341200101

[31] S. Dey, T. Mukhopadhyay, and S. Adhikari, *Uncertainty quantification in laminated composites: A meta-model based approach.,* CRC Press, 2018.

http://dx.doi.org/10.1201/9781315155593

Fuzzy Reliability Calculation Based on Stochastic Finite Element

Abstract: Based on the stochastic finite element, the fuzzy reliability calculation of structure with static problems, linear vibration, nonlinear problems and nonlinear vibration is studied. The mean and variance of stress are obtained by the stochastic finite element method. The normal membership function is generally selected as the membership function of engineering problems. The fuzzy reliability of structure can be obtained by using the calculation formula of fuzzy reliability.

Keywords: Fuzzy reliability, Stochastic finite element, Static problem, Linear vibration, Nonlinear structure, Nonlinear vibration, Membership function.

INTRODUCTION

Stress is related to load, structure and other factors. As long as one of the factors is fuzzy, the stress is fuzzy. The strength value can be found in the design manual, but this information has a certain degree of fuzziness. Finite element analysis of complex structures is feasible. Based on the stochastic finite element method, a new calculation method of fuzzy reliability of structures is proposed. This method can calculate the fuzzy reliability of complex structures.

A methodology is developed that uses Petri nets and fuzzy Lambda–Tau methodology and solves for reliability [1]. This paper presents an approach to assessing the reliability of concrete structures [2]. This paper discusses the optimisation of the forming load path using a fuzzy logic control algorithm and finite element analysis [3]. Modelling of plane strain compression (PSC) test incorporating both the hybrid and the fuzzy finite element models have been undertaken [4].This work introduces a fuzzy finite element procedure to calculate frequency response functions of damped finite element models [5]. A fuzzy logic method for improving the convergence in nonlinear magnetostatic problems using finite elements is presented [6]. This paper presents an efficient method for the static design of imprecise structures including fuzzy data [7]. In this paper, the spectral element method and a fuzzy set is used to estimate frequency response function envelopes [8]. This paper uses the interval and fuzzy finite element method for

Wenhui Mo

dynamic analysis of finite elements with uncertain parameters [9]. A methodology of the fuzzy finite element method in the model updating welded joints has been highlighted [10]. Fuzzy logic, neural networks and three-dimensional finite element calculations are employed in order to develop a computerized model in a coalface longwall mining simulation [11]. A fuzzy model of the generator is developed using finite element and fuzzy methods for carrying out its leakage field analysis [12]. Fuzzy finite element methods are becoming increasingly popular for the analysis of structure [13]. A fuzzy finite element analysis based on the α-cuts method analyzed heat conduction problems with uncertain parameters [14]. Dynamical relaxed directional method for fuzzy reliability analysis is proposed for engineering problems under epistemic uncertainty [15]. This paper describes a fuzzy reliability method for the calculation of power systems [16]. A new algorithm has been introduced to construct the membership function of fuzzy system reliability using different types of intuitionistic fuzzy failure rates [17]. A new approach of the weakest t-norm based intuitionistic fuzzy fault-tree is proposed to evaluate system reliability [18]. The objective of this study is to develop a fuzzy reliability algorithm of basic events of fault trees through qualitative data processing [19]. This study is to analyze the fuzzy reliability of a repairable system using soft-computing based hybridized techniques [20]. The stress–strength reliability of fuzziness is investigated [21]. An approach to analyze the fuzzy reliability of a dual-fuel steam turbine mechanical propulsion system is presented [22]. The purpose of the present study is to analyse the fuzzy reliability analysis of DFSMC systems with different membership functions by applying-the fuzzy lambda-tau technique [23]. A novel approach is proposed to evaluate system failure probability using intuitionistic fuzzy fault tree analysis [24]. Fuzzy improved distribution function is introduced and the order statistics based on fuzzy improved distribution function is proposed [25]. A fuzzy maximum entropy approach is proposed to determine fuzzy reliability centered maintenance considering the uncertainty [26].

Based on the stochastic finite element method, a new method for calculating the fuzzy reliability of structures with linear problems, linear vibration, nonlinear problems and nonlinear vibration is proposed. The normal membership function is selected as the membership function. The calculation formulas are given respectively.

Fuzzy Reliability Calculation of Static Problems Based on Stochastic Finite Element

The governing equation of the finite element under static load can be written as

$$[K]\{\delta\} = \{F\} \tag{1}$$

The mean and variance of structural displacement are obtained by Taylor stochastic finite element.

$$E\{\delta\} \approx \{\delta\}\big|_{a=\bar{a}} + \frac{1}{2}\sum_{i=1}^{n}\sum_{j=1}^{n}\frac{\partial^2\{\delta\}}{\partial a_i \partial a_j}\Big|_{a=\bar{a}}Cov\left(a_i,a_j\right) \tag{2}$$

where $E\{\delta\}$ is mean value $\{\delta\}$ and $Cov\left(a_i,a_j\right)$ is the covariance between a_i and a_j.

The variance of $\{\delta\}$ is given by

$$Var\{\delta\} \approx \sum_{i=1}^{n}\sum_{j=1}^{n}\frac{\partial\{\delta\}}{\partial a_i}\Big|_{a=\bar{a}} \cdot \frac{\partial\{\delta\}}{\partial a_j}\Big|_{a=\bar{a}} \cdot Cov\left(a_i,a_j\right) \tag{3}$$

where $Var\{\delta\}$ is the variance of $\{\delta\}$.

The mean and variance of the stress on the structure are obtained by Taylor stochastic finite element method

$$E\{\sigma\} \approx \{\sigma\}\big|_{a=\bar{a}} + \frac{1}{2}\sum_{i=1}^{n}\sum_{j=1}^{n}\frac{\partial^2\{\sigma\}}{\partial a_i \partial a_j}\Big|_{a=\bar{a}}Cov\left(a_i,a_j\right) \tag{4}$$

$$Var\{\sigma\} \approx \sum_{i=1}^{n}\sum_{j=1}^{n}\frac{\partial\{\sigma\}}{\partial a_i}\Big|_{a=\bar{a}} \cdot \frac{\partial\{\sigma\}}{\partial a_j}\Big|_{a=\bar{a}} \cdot Cov(a_i, a_j)$$

(5)

where, $E\{\delta\}$ is the mean of σ and $Cov(a_i, a_j)$ is the covariance between a_i and a_j. $Var\{\sigma\}$ is the variance of $\{\sigma\}$.

If the numerical range $(\sigma_{x\min}, \sigma_{x\max})$ of material strength is given in the design manual, its mean value and standard deviation are

$$n_{\sigma_x} = \frac{1}{2}(\sigma_{x\min} + \sigma_{x\max})$$

(6)

$$m_{\sigma_x} = \frac{1}{6}(\sigma_{x\max} - \sigma_{x\min})$$

(7)

The mean and variance of the stress on the structure are obtained by Neumann stochastic finite element. The geometric parameters, material parameters and load of structures are generated by a computer program.

$\{\delta\}$ is represented by the following series as

$$\{\delta\} = \{\delta_0\} - \{\delta_1\} + \{\delta_2\} - \{\delta_3\} + \cdots$$

(8)

The solution of this series is equal to the following recursive equation

$$[K_0]\{\delta_i\} = \Delta[K]\{\delta_{i-1}\}, \quad i = 1, 2, \cdots$$

(9)

Vectors $\{\delta\}_1, \{\delta\}_2, \cdots, \{\delta\}_{N_1}$ are solutions of N_1 Eqs.8.

The mean of $\{\delta\}$ is given by

$$\mu\{\delta\} = \frac{\{\delta\}_1 + \{\delta\}_2 + \cdots + \{\delta\}_{N_1}}{N_1}$$

(10)

The variance of $\{\delta\}$ is given by

$$Var\{\delta\} = \frac{1}{N_1 - 1} \sum_{i=1}^{N_1} \left(\{\delta\}_i - \mu\{\delta\}\right)^2$$

(11)

The stress of element d

$$\{\sigma\} = [D][B]\{\delta\}^d$$

(12)

The mean of stress is

$$\mu\{\sigma\} = \frac{\{\sigma\}_1 + \{\sigma\}_2 + \cdots + \{\sigma\}_{N_1}}{N_1}$$

(13)

The stress variance is

$$Var\{\sigma\} = \frac{1}{N_1 - 1} \sum_{i=1}^{N_1} \left(\{\sigma\}_i - \mu\{\sigma\}\right)^2$$

(14)

The mean and variance of the stress on the structure are obtained by the stochastic finite element using the relaxation iteration method.

The geometric parameters, material parameters and load of structures are generated by a computer program.

Successive over Relaxation iterative method solve Eq.1 as follows:

$$\delta^0 = \left\{ \{\delta\}_1^0, \cdots, \{\delta\}_n^0 \right\}^T$$

(15)

$$\{\delta\}_i^{k+1} = \{\delta\}_i^k + \omega \left(\{F\}_i - \sum_{j=1}^{i-1} [K]_{ij} \{\delta\}_j^{k+1} - \sum_{j=i}^{n} [K]_{ij} \{\delta\}_j^k \right) \Big/ [K]_{ii}$$

$$\text{(16)}$$

$$i = 1, 2, \cdots, n; k = 0, 1, \cdots$$

Vectors $\{\delta\}_1, \{\delta\}_2, \cdots, \{\delta\}_{N_1}$ are solutions of N_1 Eqs.1.

The mean of $\{\delta\}$ is given by

$$\mu\{\delta\} = \frac{\{\delta\}_1 + \{\delta\}_2 + \cdots + \{\delta\}_{N_1}}{N_1}$$

$$\text{(17)}$$

The variance of $\{\delta\}$ is given by

$$Var\{\delta\} = \frac{1}{N_1 - 1} \sum_{i=1}^{N_1} \left(\{\delta\}_i - \mu\{\delta\} \right)^2$$

$$\text{(18)}$$

The stress of element d

$$\{\sigma\} = [D][B]\{\delta\}^d$$

The mean of stress is

$$\mu\{\sigma\} = \frac{\{\sigma\}_1 + \{\sigma\}_2 + \cdots + \{\sigma\}_{N_1}}{N_1}$$

$$\text{(19)}$$

The stress variance is

$$Var\{\sigma\} = \frac{1}{N_1 - 1} \sum_{i=1}^{N_1} \left(\{\sigma\}_i - \mu\{\sigma\} \right)^2$$

$$\text{(20)}$$

Definition let X be a continuous random variable, $P(x)$ be its probability density function, and the fuzzy set on Ω represents a fuzzy event, then the probability of the fuzzy event is defined as[27]

$$P(\underset{\sim}{A}) = \int_{-\infty}^{\infty} n_{\underset{\sim}{A}}(x)P(x)\mathrm{d}x \tag{21}$$

where $n_{\underset{\sim}{A}}(x)$ is the membership function of $\underset{\sim}{A}$

The distribution of strength and working stress conform to the normal distribution: they are $r = N(\mu_r, \sigma_r^2)$, $t = N(\mu_t, \sigma_t^2)$.

μ_r and μ_t are the mathematical expectations of r and t respectively, and σ_r and σ_t are the standard deviations of r and t respectively. Let $y = r - t$, r and t are independent of each other, then y also obeys the normal distribution, and its mathematical expectation and standard deviation are

$$\mu_y = \mu_r - \mu_t \tag{22}$$

$$\sigma_y = \sqrt{\sigma_r^2 + \sigma_t^2} \tag{23}$$

The membership function of general engineering problems selects fuzzy normal distribution, and the membership function of fuzzy interference set adopts the following form

$$n_{\underset{\sim}{Y}}(y) = e^{-k(y-a)^2} \qquad (k > 0) \tag{24}$$

The membership function represents the complete information of the fuzzy set, and the failure probability of the fuzzy event is defined as [27]

$$P_f = P_f(\underset{\sim}{Y}) = \frac{1}{\sqrt{2\pi}\sigma_y} \int_{-\infty}^{0} e^{-k(y-a)^2} e^{-\frac{(y-\mu_y)^2}{2\sigma_y^2}} dy$$

$$= \frac{e^{\frac{2ak\mu_y - a^2k - k\mu_y^2}{1+2k\sigma_y^2}}}{\sqrt{1+2k\sigma_y^2}} \cdot \Phi\left(-\frac{\mu_y + 2ak\sigma_y^2}{\sigma_y \sqrt{1+2k\sigma_y^2}}\right)$$

(25)

$$= \frac{e^{\frac{2ak(\mu_r-\mu_t)-a^2k-k(\mu_r-\mu_t)^2}{1+2k(\sigma_r^2+\sigma_t^2)}}}{\sqrt{1+2k(\sigma_r^2+\sigma_t^2)}} \cdot \Phi\left(-\frac{\mu_r - \mu_t + 2ak(\sigma_r^2 + \sigma_t^2)}{\sqrt{\sigma_r^2 + \sigma_t^2 + 2k(\sigma_r^2 + \sigma_t^2)^2}}\right)$$

The fuzzy reliability of the strength of the structure is

$$R = 1 - P_f \tag{26}$$

Fuzzy Reliability Calculation of Structures with Linear Vibration

For a linear system, the dynamic equilibrium equation is given by

$$[M]\{\ddot{\delta}\} + [C]\{\dot{\delta}\} + [K]\{\delta\} = \{F\} \tag{27}$$

where $[M],[K]$ and $[C]$ are the global mass, stiffness and damping matrices.

See Chapter 2 for the solution of mean and variance of stress. The fuzzy reliability of the strength of the structure with linear vibration can be obtained by using Eqs.25, 26.

Fuzzy Reliability of Nonlinear Structures

See chapter 1 or chapter 2 for the solution of mean and variance of stress. The fuzzy reliability of the strength of the structure with nonlinear structure can be obtained by using Eqs.25, 26.

Fuzzy Reliability of Structures with Nonlinear Vibration

See chapter 2 for the solution of mean and variance of stress. The fuzzy reliability of the strength of the structure with nonlinear structure can be obtained by using Eqs.25, 26.

CONCLUDING REMARKS

Considering the influence of fuzzy factors on the structure, the fuzzy reliability of the structure should be calculated. The calculation formulas of fuzzy reliability of structure with linear problems, linear vibration, nonlinear problems and nonlinear vibration are proposed. The method proposed in this chapter can be used to calculate the fuzzy reliability of complex structures.

REFERENCES

[1] J. Knezevic, and E.R. Odoom, "Reliability modelling of repairable systems using Petri nets and fuzzy Lambda–Tau methodology", *Reliab. Eng. Syst. Saf.,* vol. 73, no. 1, pp. 1-17, 2001.
      http://dx.doi.org/10.1016/S0951-8320(01)00017-5

[2] F. Biondini, F. Bontempi, and P. Giorgio Malerba, "FrancoBontempi,PierGiorgio Malerba, "Fuzzy reliability analysis of concrete structures", *Comput. Struc.,* vol. 82, no. 13-14, pp. 1033-1052, 2004.
      http://dx.doi.org/10.1016/j.compstruc.2004.03.011

[3] P. Ray, and B.J. Mac Donald, "Determination of the optimal load path for tube hydroforming processes using a fuzzy load control algorithm and finite element analysis", *Finite Elem. Anal. Des.,* vol. 41, no. 2, pp. 173-192, 2004.
      http://dx.doi.org/10.1016/j.finel.2004.03.005

[4] M.F. Abbod, J. Talamantes-Silva, D.A. Linkens, and I. Howard, JTalamantes-Silva,D.ALinkens, I Howard, "Modelling of plane strain compression (PSC) test for aluminium alloys using finite elements and fuzzy logic", *Eng. Appl. Artif. Intell.,* vol. 17, no. 5, pp. 447-456, 2004.
      http://dx.doi.org/10.1016/j.engappai.2004.04.001

[5] H. De Gersem, D. Moens, W. Desmet, and D. Vandepitte, "A fuzzy finite element procedure for the calculation of uncertain frequency response functions of damped structures: Part 2—Numerical case studies", *J. Sound Vibrat.,* vol. 288, no. 3, pp. 463-486, 2005.
      http://dx.doi.org/10.1016/j.jsv.2005.07.002

[6]   M.A. Arjona, L.,R.Escarela-Perez,E.Melgoza-Vázquez, "Convergence improvement in two-dimensional finite element nonlinear magnetic problems—a fuzzy logic approach", *Finite Elem. Anal. Des.,* vol. 41, no. 6, pp. 583-598, 2005.
      http://dx.doi.org/10.1016/j.finel.2004.10.002

[7]   F. Massa, T. Tison, and B. Lallemand, "A fuzzy procedure for the static design of imprecise structures", *Comput. Methods Appl. Mech. Eng.,* vol. 195, no. 9-12, pp. 925-941, 2006.
      http://dx.doi.org/10.1016/j.cma.2005.02.015

[8]   R.F. Nunes, A. Klimke, and J.R.F. Arruda, "On estimating frequency response function envelopes using the spectral element method and fuzzy sets", *J. Sound Vibrat.,* vol. 291, no. 3-5, pp. 986-1003, 2006.
      http://dx.doi.org/10.1016/j.jsv.2005.07.024

[9]   H. De Gersem, D. Moens, W. Desmet, and D. Vandepitte, DavidMoens,WimDesmet,DirkVandepitte, "Interval and fuzzy dynamic analysis of finite element models with superelements", *Comput. Struc.,* vol. 85, no. 5-6, pp. 304-319, 2007.
      http://dx.doi.org/10.1016/j.compstruc.2006.10.011

[10] O. Ait-Salem Duque, A.R. Senin, A. Stenti, M. De Munck, and F. Aparicio, "A methodology for the choice of the initial conditions in the model updating of welded joints using the fuzzy finite element method", *Comput. Struc.,* vol. 85, no. 19-20, pp. 1534-1546, 2007.
      http://dx.doi.org/10.1016/j.compstruc.2007.01.016

[11] J. Toraño, I. Diego, M. Menéndez, and M. Gent, "IsidroDiego,MarioMenéndez,MalcolmGent, "A finite element method (FEM) – Fuzzy logic (Soft Computing) – virtual reality model approach in a coalface longwall mining simulation", *Autom. Construct.,* vol. 17, no. 4, pp. 413-424, 2008.
      http://dx.doi.org/10.1016/j.autcon.2007.07.001

[12] A. Kumar, S. Marwaha, A. Singh, and A. Marwaha, "SanjayMarwaha,AmarpalSingh,AnupamaMarwaha, "Comparative leakage field analysis of electromagnetic devices using finite element and fuzzy methods", *Expert Syst. Appl.,* vol. 37, no. 5, pp. 3827-3834, 2010.
      http://dx.doi.org/10.1016/j.eswa.2009.11.036

[13] L. Farkas, D. Moens, D. Vandepitte, and W. Desmet, "Fuzzy finite element analysis based on reanalysis technique", *Struct. Saf.,* vol. 32, no. 6, pp. 442-448, 2010.
      http://dx.doi.org/10.1016/j.strusafe.2010.04.004

[14] M. Bart, "Nicolaï,Jose A.Egea,NicoScheerlinck,Julio R.Banga,Ashim K.Datta, "Fuzzy finite element analysis of heat conduction problems with uncertain parameters", *J. Food Eng.,* vol. 103, no. 1, pp. 38-46, 2011.
      http://dx.doi.org/10.1016/j.jfoodeng.2010.09.017

[15] M. Bagheri, S.A. Hosseini, and B. Keshtegar, "Seyed AbbasHosseini,BehroozKeshtegar, "Dynamical relaxed directional method for fuzzy reliability analysis", *Structures,* vol. 34, pp. 169-179, 2021.
      http://dx.doi.org/10.1016/j.istruc.2021.07.043

[16] S.S. Halilčević, F. Gubina, and A.F. Gubina, "FerdinandGubina,Andrej FerdoGubina, "The composite fuzzy reliability index of power systems", *Eng. Appl. Artif. Intell.,* vol. 24, no. 6, pp. 1026-1034, 2011.
      http://dx.doi.org/10.1016/j.engappai.2011.04.009

[17] M. Kumar, and S.P. Yadav, "A novel approach for analyzing fuzzy system reliability using different types of intuitionistic fuzzy failure rates of components", *ISA Trans.,* vol. 51, no. 2, pp. 288-297, 2012.
      http://dx.doi.org/10.1016/j.isatra.2011.10.002 PMID: 22134065

[18] M. Kumar, and S.P. Yadav, "The weakest t-norm based intuitionistic fuzzy fault-tree analysis to evaluate system reliability", *ISA Trans.,* vol. 51, no. 4, pp. 531-538, 2012.

http://dx.doi.org/10.1016/j.isatra.2012.01.004 PMID: 22445394

[19] J.H. Purba, J. Lu, G. Zhang, and W. Pedrycz, JieLu,GuangquanZhang,WitoldPedrycz,"A fuzzy reliability assessment of basic events of fault trees through qualitative data processing", *Fuzzy Sets Syst.,* vol. 243, pp. 50-69, 2014.

http://dx.doi.org/10.1016/j.fss.2013.06.009

[20] S.P. Komal, and S.P. Sharma, Sharma, "Fuzzy reliability analysis of repairable industrial systems using soft-computing based hybridized techniques", *Appl. Soft Comput.,* vol. 24, pp. 264-276, 2014.

http://dx.doi.org/10.1016/j.asoc.2014.06.054

[21] S. Eryilmaz, and G.Y. Tütüncü, G. YazgıTütüncü, "Stress strength reliability in the presence of fuzziness", *J. Comput. Appl. Math.,* vol. 282, pp. 262-267, 2015.

http://dx.doi.org/10.1016/j.cam.2014.12.047

[22] Komal, D. Chang, and S. Lee, Komal,DaejunChang,Seong-yeobLee, "Fuzzy reliability analysis of dual-fuel steam turbine propulsion system in LNG carriers considering data uncertainty", *J. Nat. Gas Sci. Eng.,* vol. 23, pp. 148-164, 2015.

http://dx.doi.org/10.1016/j.jngse.2015.01.030

[23] Komal, Komal, "Fuzzy reliability analysis of DFSMC system in LNG carriers for components with different membership function", *Ocean Eng.,* vol. 155, pp. 278-294, 2018.

http://dx.doi.org/10.1016/j.oceaneng.2018.02.061

[24] M. Kumar, and M. Kaushik, ManviKaushik, "System failure probability evaluation using fault tree analysis and expert opinions in intuitionistic fuzzy environment", *J. Loss Prev. Process Ind.,* vol. 67, 2020.104236

http://dx.doi.org/10.1016/j.jlp.2020.104236

[25] I. Bayramoglu, "Fuzzy improved distribution function and order statist- ics", *J. Comput. Appl. Math.,* vol. 411, no. September, 2022.114266

http://dx.doi.org/10.1016/j.cam.2022.114266

[26] A. Marco, Fuentes-Huerta, David S.González-González, MarioCantú-Sifuentes, Rolando J.Praga-Alejo, "Reliability centered maintenance considering personnel experience and only censored data", *Comput. Ind. Eng.,* vol. 158, 2021.107440

http://dx.doi.org/10.1016/j.cie.2021.107440

[27] L.A. Zadeh, "Probability measures of fuzzy events", *J. Math. Anal. Appl.,* vol. 23, no. October, pp. 421-427, 1968.

http://dx.doi.org/10.1016/0022-247X(68)90078-4

CHAPTER 4

Static Analysis of Interval Finite Element

Abstract: Four methods of interval finite element for static analysis are proposed. Using the second-order and third-order Taylor expansion , interval finite element for static analysis is addressed. Neumann expansion of interval finite element for static analysis is formulated. Interval finite element using Sherman -Morrison-Woodbury expansion is presented. A new iterative method (NIM) is used for interval finite element calculation. Four methods can calculate the upper and lower bounds of node displacement and element stress.

Keywords: A new iterative method (NIM), Interval finite element, Neumann expansion, Taylor expansion, Sherman-Morrison-Woodbury expansion, Static analysis.

INTRODUCTION

Final element method deals with deterministic engineering problems has become a world recognized numerical analysis method. Stochastic finite element has been developed to analyze structure with stochastic parameters. Fuzzy finite element has been developed to analyze structure with fuzzy parameters. In the design of engineering problems, material properties, geometry parameters and loads are assumed to be interval variables. Instead of conventional finite elements, the interval finite element has been studied by many authors.

This work analyzes structural systems using interval analysis [1]. A formulation is proposed for the interval estimation of displacement input with uncertainty [2].Interval calculation is used to analyze mechanical systems modeled with interval finite elements [3]. The uncertain parameters are assumed to be interval variables, and the bounds of the displacement are obtained by interval finite element methods [4]. To account for uncertainties in linear static problems, a interval linear equations is proposed [5]. The interval and fuzzy finite element method is used to analyze the eigenvalue and frequency response function analysis of structures [6]. This paper presents a method for computing linear systems with large uncertainties

[7].Affine arithmetic is an improving interval analysis in finite element calculations [8].This paper presents chosen computational algorithms for interval finite element analysis [9]. The article focuses on a new interval finite element formulation to reduce overestimation problems [10]. This study is deemed a contribution to novel parameterized intervals for solving problems with uncertainty [11]. Improved interval analysis of the second-order statistics of the response is proposed [12]. Optimization and anti-optimization solution of combined parameterized and improved interval analyses for structures with uncertainties is presented [13].The objective is to validate a methodology for multivariate uncertainty in interval finite elements [14]. Interval Finite element analysis of linear-elastic structures with uncertain properties is addressed [15]. This paper presents a flexible approach for intervalfinite element analysis [16].

Four static analysis methods of interval finite element are proposed. They are Taylor expansion method, Neumann expansion method, Sherman-Morrison-Woodbury expansion method and a new iterative method (NIM). The calculation process and formulas of the four methods are given in detail

Taylor Expansion for Interval Finite Element

Interval variable $\left[\underline{a}, \bar{a}\right]$ is generated by the following formula

$$a_i = \underline{a} + \frac{\bar{a} - \underline{a}}{n} i = \frac{i\bar{a} + (n-i)\underline{a}}{n} \tag{1}$$

$$i = 1, 2, \cdots, n$$

Material properties, geometry parameters and applied loads of structures are assumed to be interval variables. They are $[\underline{a}_1, \bar{a}_1], [\underline{a}_2, \bar{a}_2], \cdots, [\underline{a}_j, \bar{a}_j], \cdots, [\underline{a}_n, \bar{a}_n]$.

The equilibrium equation is written as

$$KU = F \tag{2}$$

where U = the displacement vector, F = the external force, K = the global stiffness matrix.

By applying Taylor series at the midpoint of the interval variables, the following equations are given by

$$U^0 = \left(K^0\right)^{-1} F^0 \tag{3}$$

$$\frac{\partial U}{\partial a_i} = \left(K^0\right)^{-1}\left(\frac{\partial F}{\partial a_i} - \frac{\partial K}{\partial a_i} U^0\right) \tag{4}$$

$$\frac{\partial^2 U}{\partial a_i \partial a_j} = \left(K^0\right)^{-1}\left(\frac{\partial^2 F}{\partial a_i \partial a_j} - \frac{\partial^2 K}{\partial a_i \partial a_j} U^0 - \frac{\partial K}{\partial a_i}\frac{\partial U}{\partial a_j} - \frac{\partial K}{\partial a_j}\frac{\partial U}{\partial a_i}\right) \tag{5}$$

$$\frac{\partial^3 U}{\partial a_i^2 \partial a_j} = (K^0)^{-1}\left(\frac{\partial^3 F}{\partial a_i^2 \partial a_j} - \frac{\partial^3 K}{\partial a_i^2 \partial a_j} U^0 - 3\frac{\partial^2 K}{\partial a_i \partial a_j}\frac{\partial U}{\partial a_i} - 3\frac{\partial K}{a_i}\frac{\partial^2 U}{\partial a_i \partial a_j}\right) \tag{6}$$

The Taylor expansion formula of U is

$$U = U^0 + \sum_{k=1}^{m}\frac{1}{k!}\sum_{i_1,i_2,\cdots,i_k=1}^{n}\frac{\partial^k U}{\partial a_{i_1}\partial a_{i_2}\cdots\partial a_{i_k}}\left(a^0\right)\left(a_{i_1} - a^0_{i_1}\right)$$

$$\left(a_{i_2} - a^0_{i_2}\right)\cdots\left(a_{i_k} - a^0_{i_k}\right) +$$

$$\tag{7}$$

$$\frac{1}{(m+1)!}\sum_{i_1,i_2,\cdots,i_{m+1}=1}^{n}\frac{\partial^{m+1} U}{\partial a_{i_1}\partial a_{i_2}\cdots\partial a_{i_k}}\left(a^0\right)\left(a_{i_1} - a^0_{i_1}\right)$$

$$\left(a_{i_2} - a^0_{i_2}\right)\cdots\left(a_{i_k} - a^0_{i_k}\right)$$

The second-order term of the Taylor expansion formula is given by

$$U \approx U^0 + \sum_{k=1}^{2} \frac{1}{k!} \sum_{i_1, i_2, \cdots, i_k = 1}^{n} \frac{\partial^k U}{\partial a_{i_1} \partial a_{i_2} \cdots \partial a_{i_k}} \left(a^0 \right) \left(a_{i_1} - a^0_{i_1} \right)$$

$$\left(a_{i_2} - a^0_{i_2} \right) \cdots \left(a_{i_k} - a^0_{i_k} \right)$$

(8)

The third-order Taylor expansion formula of U is

$$U \approx U^0 + \sum_{k=1}^{3} \frac{1}{k!} \sum_{i_1, i_2, \cdots, i_k = 1}^{n} \frac{\partial^k U}{\partial a_{i_1} \partial a_{i_2} \cdots \partial a_{i_k}} \left(a^0 \right) \left(a_{i_1} - a^0_{i_1} \right)$$

$$\left(a_{i_2} - a^0_{i_2} \right) \cdots \left(a_{i_k} - a^0_{i_k} \right)$$

(9)

Substituting N_1 samples of interval variables into Eq.8 or Eq.9, the vectors $U_1, U_2, \cdots, U_{N_1}$ can be obtained.

The maximums and minimums of components of $U_1, U_2, \cdots, U_{N_1}$ are the upper bounds and lower bounds.

$$Upper_x = \max(U_{1x}, U_{2x}, \cdots, U_{N_1 x})$$

(10)

where max ( ) is the maximum value of the x-direction component of N_1 displacement vectors, $Upper_x$ is the upper value of the x-direction component of N_1 displacement vectors.

$$Lower_x = \min(U_{1x}, U_{2x}, \cdots, U_{N_1 x})$$

(11)

where min( ) is the minimum value of the x-direction component of N_1 displacement vectors, $Lower_x$ is the lower value of the x-direction component of N_1 displacement vectors.

$$Upper_y = \max(U_{1y}, U_{2y}, \cdots, U_{N_1 y})$$ **(12)**

where max( ) is the maximum value of the y-direction component of N_1 displacement vectors, $Upper_y$ is the upper value of the y-direction component of N_1 displacement vectors.

$$Lower_y = \min(U_{1y}, U_{2y}, \cdots, U_{N_1 y})$$ **(13)**

where min( ) is the minimum value of the y-direction component of N_1 displacement vectors, $Lower_y$ is the upper value of the y-direction component of N_1 displacement vectors.

$$Upper_z = \max(U_{1z}, U_{2z}, \cdots, U_{N_1 z})$$ **(14)**

where max( ) is the maximum value of the z-direction component of N_1 displacement vectors, $Upper_z$ is the upper value of the z-direction component of N_1 displacement vectors.

$$Lower_z = \min(U_{1z}, U_{2z}, \cdots, U_{N_1 z})$$ **(15)**

where min( ) is the minimum value of the z-direction component of N_1 displacement vectors, $Lower_z$ is the lower value of the z-direction component of N_1 displacement vectors.

The strain and stress vectors for an element d are

$$\{\varepsilon\} = [B]U^d$$ **(16)**

and

$$\{\sigma\} = [D]\{\varepsilon\}$$ **(17)**

where, $[D]$ =the material response matrix of an element d, $[B]$ = thegradient matrix of element d and U^d = the element d nodal displacement vector .

$$\{\sigma\} = [D][B]U \tag{18}$$

The stress for an element d is given by

$$\{\sigma\} = [D][B]U^d \tag{19}$$

Substituting $N_1 U^d$ and N_1 samples of interval variables into Eq.19, the vectors $\{\sigma\}_1, \{\sigma\}_2, \cdots, \{\sigma\}_{N_1}$ can be obtained.

$$Upper_{\sigma x} = \max(\sigma_{1x}, \sigma_{2x}, \cdots, \sigma_{N_1 x}) \tag{20}$$

where max( ) is the maximum value of the x-direction component of N_1 stress vectors, $Upper_{\sigma x}$ is the upper value of the x-direction component of N_1 stress vectors.

$$Lower_{\sigma x} = \min(\sigma_{1x}, \sigma_{2x}, \cdots, \sigma_{N_1 x}) \tag{21}$$

where min( ) is the minimum value of the x-direction component of N_1 stress vectors, $Lower_{\sigma x}$ is the lower value of the x-direction component of N_1 stress vectors.

$$Upper_{\sigma y} = \max(\sigma_{1y}, \sigma_{2y}, \cdots, \sigma_{N_1 y}) \tag{22}$$

where max( ) is the maximum value of the y-direction component of N_1 stress vectors, $Upper_{\sigma y}$ is the upper value of the y-direction component of N_1 stress vectors.

$$Lower_{\sigma y} = \min(\sigma_{1y}, \sigma_{2y}, \cdots, \sigma_{N_1 y}) \tag{23}$$

where min( ) is the minimum value of the y-direction component of N_1 stress vectors, *Lower*$_{\sigma y}$ is the lower value of the y-direction component of N_1 stress vectors.

$$Upper_{\sigma z} = \max(\sigma_{1z}, \sigma_{2z}, \cdots, \sigma_{N_1 z}) \tag{24}$$

where max( ) is the maximum value of the z-direction component of N_1 stress vectors, *Upper*$_{\sigma z}$ is the upper value of the z-direction component of N_1 stress vectors.

$$Lower_{\sigma z} = \min(\sigma_{1z}, \sigma_{2z}, \cdots, \sigma_{N_1 z}) \tag{25}$$

where min( ) is the minimum value of the z-direction component of N_1 stress vectors, *Lower*$_{\sigma z}$ is the lower value of the z-direction component of N_1 stress vectors.

$$Upper_{\tau xy} = \max(\tau_{1xy}, \tau_{2xy}, \cdots, \tau_{N_1 xy}) \tag{26}$$

where max( ) is the maximum value of N_1 shear stresses in xy plane, *Upper*$_{\tau xy}$ is the upper value of N_1 shear stresses in xy plane.

$$Lower_{\tau xy} = \min(\tau_{1xy}, \tau_{2xy}, \cdots, \tau_{N_1 xy}) \tag{27}$$

where min( ) is the minimum value of N_1 shear stresses in xy plane, *Lower*$_{\tau xy}$ is the lower value of N_1 shear stresses in xy plane.

$$Upper_{\tau yz} = \max(\tau_{1yz}, \tau_{2yz}, \cdots, \tau_{N_1 yz}) \tag{28}$$

where max( ) is the maximum value of N_1 shear stresses in yz plane, *Upper*$_{\tau yz}$ is the upper value of N_1 shear stresses in yz plane.

$$Lower_{\tau yz} = \min(\tau_{1yz}, \tau_{2yz}, \cdots, \tau_{N_1 yz}) \tag{29}$$

where min( ) is the minimum value of N_1 shear stresses in yz plane, *Lower*$_{\tau yz}$ is the lower value of N_1 shear stresses in yz plane.

$$Upper_{\tau zx} = \max(\tau_{1zx}, \tau_{2zx}, \cdots, \tau_{N_1zx}) \tag{30}$$

where max( ) is the maximum value of N_1 shear stresses in zx plane, $Upper_{\tau zx}$ is the upper value of N_1 shear stresses in zx plane.

$$Lower_{\tau zx} = \min(\tau_{1zx}, \tau_{2zx}, \cdots, \tau_{N_1zx}) \tag{31}$$

where min( ) is the minimum value of N_1 shear stresses in zx plane, $Lower_{\tau zx}$ is the lower value of N_1 shear stresses in zx plane.

Interval Finite Element using Neumann Expansion

Neumann expansion is applied to stochastic finite elements [17][18]. Neumann expansion is also applied to static analysis of interval finite elements. Interval variable $\left[\underline{a}, \overline{a}\right]$ is generated by the following formula

$$a_i = \underline{a} + \frac{\overline{a}-\underline{a}}{n}i - \frac{\overline{a}-\underline{a}}{2} = \frac{(3n-2i)\underline{a}+(2i-n)\overline{a}}{2n} \tag{32}$$

$$i = 1, 2, \cdots, n$$

Material properties, geometry parameters and applied loads of structures are assumed to be interval variables.

The equilibrium equation is written as

$$KU = F \tag{33}$$

where U = the displacement vector, F = the external force, K = the global stiffness matrix.

The stiffness matrix K is decomposed into two matrices

$$K = K_0 + \Delta K \tag{34}$$

where K_0 is the stiffness matrix is replaced by mean values, ΔK representing deviatoric parts. The solution U_0 can be obtained as

$$U_0 = K_0^{-1}F \tag{35}$$

The Neumann expansion of K^{-1} takes the following form:

$$K^{-1} = (K_0 + \Delta K)^{-1} = (I - P + P^2 - P^3 + L)K_0^{-1} \tag{36}$$

U is represented by the following series as

$$U = U_{(0)} - U_{(1)} + U_{(2)} - U_{(3)} + \cdots \tag{37}$$

This series solution is equivalent to the following equation:

$$K_0 U_{(i)} = \Delta K U_{(i-1)} \quad i = 1, 2, \ldots \tag{38}$$

It may be terminated if the following criterion is confirmed.

$$\frac{\left\|U_i\right\|_2}{\left\|\sum_{k=0}^{i}(-1)^k U_{(k)}\right\|_2} \le \delta_{err} \tag{39}$$

where δerr is the allowable error and $\|\cdot\|_2$ is the vector norm defined by

$$\|U\|_2 = \sqrt{U^T U} \tag{40}$$

It is well known that the Neumann expansion converges if the absolute values of all the eigenvalues $K_0^{-1}\Delta K$ are less than 1.

Substituting N_1 samples of interval variables into above formulas, the vectors $U_1, U_2, \cdots, U_{N_1}$ can be obtained.

The maximums and minimums of components of $U_1, U_2, \cdots, U_{N_1}$ are the upper bounds and lower bounds.

$$Upper_x = \max(U_{1x}, U_{2x}, \cdots, U_{N_1 x}) \tag{41}$$

where max( ) and $Upper_x$ see the previous section..

$$Lower_x = \min(U_{1x}, U_{2x}, \cdots, U_{N_1 x}) \tag{42}$$

where min( ) and $Lower_x$ see the previous section.

$$Upper_y = \max(U_{1y}, U_{2y}, \cdots, U_{N_1 y}) \tag{43}$$

where max( ) and $Upper_y$ see the previous section

$$Lower_y = \min(U_{1y}, U_{2y}, \cdots, U_{N_1 y}) \tag{44}$$

where min( ) and $Lower_y$ see the previous section.

$$Upper_z = \max(U_{1z}, U_{2z}, \cdots, U_{N_1 z}) \tag{45}$$

where max( ) is and $Upper_z$ see the previous section..

$$Lower_z = \min(U_{1z}, U_{2z}, \cdots, U_{N_1 z}) \tag{46}$$

where min( ) and $Lower_z$ see the previous section.

The stress for the element d is given by

$$\{\sigma\} = [D][B]U^d \tag{47}$$

Substituting $N_1 U^d$ and N_1 samples of interval variables into Eq.47, the vectors $\{\sigma\}_1, \{\sigma\}_2, \cdots, \{\sigma\}_{N_1}$ can be obtained.

$$Upper_{\sigma x} = \max(\sigma_{1x}, \sigma_{2x}, \cdots, \sigma_{N_1 x}) \tag{48}$$

where max( ) and $Upper_{\sigma x}$ see the previous section.

$$Lower_{\sigma x} = \min(\sigma_{1x}, \sigma_{2x}, \cdots, \sigma_{N_1 x}) \tag{49}$$

where min( ) and $Lower_{\sigma x}$ see the previous section.

$$Upper_{\sigma y} = \max(\sigma_{1y}, \sigma_{2y}, \cdots, \sigma_{N_1 y}) \tag{50}$$

where max( ) and $Upper_{\sigma y}$ see the previous section

$$Lower_{\sigma y} = \min(\sigma_{1y}, \sigma_{2y}, \cdots, \sigma_{N_1 y}) \tag{51}$$

where min( ) and $Lower_{\sigma y}$ see the previous section.

$$Upper_{\sigma z} = \max(\sigma_{1z}, \sigma_{2z}, \cdots, \sigma_{N_1 z}) \tag{52}$$

where max( ) and $Upper_{\sigma z}$ see the previous section.

$$Lower_{\sigma z} = \min(\sigma_{1z}, \sigma_{2z}, \cdots, \sigma_{N_1 z}) \tag{53}$$

where min( ) and $Lower_{\sigma z}$ see the previous section.

$$Upper_{\tau xy} = \max(\tau_{1xy}, \tau_{2xy}, \cdots, \tau_{N_1 xy}) \tag{54}$$

where max( ) and $Upper_{\tau xy}$ see the previous section.

$$Lower_{\tau xy} = \min(\tau_{1xy}, \tau_{2xy}, \cdots, \tau_{N_1 xy}) \tag{55}$$

where   min   ( )   and $Lower_{\tau xy}$ see   the   previous   section.

$$Upper_{\tau yz} = \max(\tau_{1yz}, \tau_{2yz}, \cdots, \tau_{N_1yz}) \tag{56}$$

where max( ) and $Upper_{\tau yz}$ see the previous section.

$$Lower_{\tau yz} = \min(\tau_{1yz}, \tau_{2yz}, \cdots, \tau_{N_1yz}) \tag{57}$$

where min( ) and $Lower_{\tau yz}$ see the previous section.

$$Upper_{\tau zx} = \max(\tau_{1zx}, \tau_{2zx}, \cdots, \tau_{N_1zx}) \tag{58}$$

where max( ) and $Upper_{\tau zx}$ see the previous section.

$$Lower_{\tau zx} = \min(\tau_{1zx}, \tau_{2zx}, \cdots, \tau_{N_1zx}) \tag{59}$$

where min( ) and $Lower_{\tau zx}$ see the previous section.

Interval Finite Element using Sherman-Morrison-Woodbury Expansion

Samples of interval variables are generated by Eq.32.

Sherman-Morrison-Woodbury expansion is expressed as

$$(K + UV^T)^{-1} = K^{-1} - K^{-1}U(I + V^T K^{-1}U)^{-1}V^T K^{-1} \tag{60}$$

We obtain

$$(K_0 + (\Delta K^T)^T)^{-1} = K_0^{-1} - K_0^{-1}\left(I + (\Delta K^T)^T K_0^{-1}\right)^{-1}(\Delta K^T)^T K_0^{-1} \tag{61}$$

Eq.2 is rewritten as

$$U = (K_0 + (\Delta K^T)^T)^{-1}F \tag{62}$$

$$= K_0{}^{-1}F - K_0{}^{-1}\left(I + (\Delta K^T)^T K_0{}^{-1}\right)^{-1}(\Delta K^T)^T K_0{}^{-1}F \qquad (63)$$

Using Neumann expansion, the above formula is $U = (K_0 + (\Delta K^T)^T)^{-1}F$

$$= K_0{}^{-1}F - K_0{}^{-1}\left(-(\Delta K^T)^T K_0{}^{-1} + \left((\Delta K^T)^T K_0{}^{-1}\right)^2\right)(\Delta K^T)^T K_0{}^{-1}F \qquad (64)$$

Substituting N_1 samples of interval variables into Eq.63 or Eq.64, the vectors $U_1, U_2, \cdots, U_{N_1}$ can be obtained.

The maximums and minimums of component of $U_1, U_2, \cdots, U_{N_1}$ are the upper bounds and lower bounds. See the above section for the calculation formula of the upper and lower bounds of displacement.

The stress for an element d is given by

$$\{\sigma\} = [D][B]\{u^d\} \qquad (65)$$

Substituting $N_1 U^d$ and N_1 samples of interval variables into Eq.65, the vectors $\{\sigma\}_1, \{\sigma\}_2, \cdots, \{\sigma\}_{N_1}$ can be obtained. The maximums and minimums of $\{\sigma\}_1, \{\sigma\}_2, \cdots, \{\sigma\}_{N_1}$ are the upper bounds and lower bounds of $\{\sigma\}_1, \{\sigma\}_2, \cdots, \{\sigma\}_{N_1}$. See the above section for the calculation formula of the upper and lower bounds of stress.

A New Iterative Method (NIM)

Material properties, geometry parameters and applied loads of structures are assumed to be interval variables. Samples of interval variables are generated by Eq.1.

Eq.2 is rewritten as

Ax=b,   (66)

In this section, a new iterative method (NIM)and its convergence analysis for finding solution of Eq.66 along with the estimation oferror bounds are described. Let A∈Cm×n and T be a subspace of Cn. Starting with Z0=βY, where β is a non zero realscalar,Y∈C$^{n\times m}$ satisfying R(Y) ⊆T and for any x0 ∈_T,the iterative method [19]is defined for k=0,1,2,... by

$$Z_{k+1}=Z_k(2I-AZ_k) \tag{67}$$

$$x_{k+1}=x_k+Z_{k+1}(b-Ax_k) \tag{68}$$

Substituting N_1 samples of interval variables into Eq.68, the vectors $U_1, U_2, \cdots, U_{N_1}$ can be obtained. The maximums and minimums of $U_1, U_2, \cdots, U_{N_1}$ are the lower bounds and upper bounds of $U_1, U_2, \cdots, U_{N_1}$.

See section 2 for the calculation formula of the upper and lower bounds of displacement.

The stress for an element d is given by

$$\{\sigma\} = [D][B]\{U^d\} \tag{69}$$

Substituting $N_1 U^d$ and N_1 samples of interval variables into Eq.69, the vectors $\{\sigma\}_1, \{\sigma\}_2, \cdots, \{\sigma\}_{N_1}$ can be obtained. The maximums and minimums of $\{\sigma\}_1, \{\sigma\}_2, \cdots, \{\sigma\}_{N_1}$ are the upper bounds and lower bounds of $\{\sigma\}_1, \{\sigma\}_2, \cdots, \{\sigma\}_{N_1}$. See section 2 for the calculation formula of the upper and lower bounds of stress.

CONCLUDING REMARKS

Considering the influence of uncertain factors on the structure, four methods of interval finite element for static calculation are proposed. The methods in this chapter are of universal significance. They can be used for the static calculation of interval finite elements such as dams, buildings, bridges and mechanical parts, *etc.*

REFERENCES

[1] S.S. Rao, and L. Berke, "Analysis of Uncertain Structural Systems Using Interval Analysis", *AIAA J.,* vol. 35, pp. 727-735, 1997.

      http://dx.doi.org/10.2514/2.164

[2] S. Nakagiri, and K. Suzuki, "Finite element interval analysis of external loads identified by displacement input with uncertainty", *Comput. Methods Appl. Mech. Eng.,* vol. 168, pp. 63-72, 1999.

      http://dx.doi.org/10.1016/S0045-7825(98)00134-0

[3] O. Dessombz, F. Thouverez, J-P. La^ın'e, and L. J'ez'eque, "Analysis of Mechanical Systems using Interval Computations applied to Finite Elements Methods", *J. Sound Vibrat.,* no. January, pp. 1-21, 2000.

[4] R.L. Muhanna, and R.L. Mullen, "Uncertainty in mechanics problems-interval-based approach", *J. Eng. Mech.,* no. JUNE, pp. 557-566, 2001.

      http://dx.doi.org/10.1061/(ASCE)0733-9399(2001)127:6(557)

[5] S. Mc William, "Anti-optimisation of uncertain structures using interval analysis", *Comput. Struc.,* no. May, pp. 421-430, 2001.

      http://dx.doi.org/10.1016/S0045-7949(00)00143-7

[6] David Moens,WimDesmet,DirkVandepitte, "Interval and fuzzy dynamic analysis of finite element models with superelements", *Comput. Struc.,* vol. 85, pp. 304-319, 2007.

      http://dx.doi.org/10.1016/j.compstruc.2006.10.011

[7] A. Neumaier, and A. Pownuk, "Linear systems with large uncertainties, with applications to truss structures", *Reliab. Comput.,* vol. 13, pp. 149-172, 2007.

      http://dx.doi.org/10.1007/s11155-006-9026-1

[8] D. Degrauwe, G. Lombaert, and G. De Roeck, "Improving interval analysis in finite element calculations by means of affine arithmetic", *Comput. Struc.,* vol. 88, pp. 247-254, 2010.

      http://dx.doi.org/10.1016/j.compstruc.2009.11.003

[9] M.V. Rama Rao, R.L. Mullen, and R.L. Muhanna, *A new interval finite element formulation with the same accuracy in primary and derived variables.*

[10] I. Elishakoff, and Y. Miglis, *"Novel parameterized intervals may lead to sharp bounds"*, Mech. Res. Comm., vol. Vol. 44, 2012, pp. 1-8.

[11] G. Muscolino, and A. Sofi, "Stochastic analysis of structures with uncertain-but-bounded parameters *via* improved interval analysis", *Probab. Eng. Mech.,* vol. 28, pp. 152-163, 2012.

      http://dx.doi.org/10.1016/j.probengmech.2011.08.011

[12] MilanVaško,PeterPecháč, "Chosen numerical algorithms for interval finite element analysis", *Procedia Eng.,* vol. 96, pp. 400-409, 2014.

      http://dx.doi.org/10.1016/j.proeng.2014.12.109

[13] R. Santoro, G. Muscolino, and I. Elishakoff, "Optimization and anti-optimization solution of combined parameterized and improved interval analyses for structures with uncertainties", *Comput. Struc.,* vol. 149, pp. 31-42, 2015.

      http://dx.doi.org/10.1016/j.compstruc.2014.11.006

[14] M. Faes, J. Cerneels, D. Vandepitte, and D. Moens, "Identification and quantification of multivariate interval uncertainty in finite element models", *Comput. Methods Appl. Mech. Eng.,* vol. 315, pp. 896-920, 2017.

      http://dx.doi.org/10.1016/j.cma.2016.11.023

[15] A. Sofi, E. Romeo, O. Barrera, and A. Cocks, "An interval finite element method for the analysis of structures with spatially varying uncertainties", *Adv. Eng. Softw.,* vol. 128, pp. 1-19, 2019.
http://dx.doi.org/10.1016/j.advengsoft.2018.11.001

[16] M. Faes, and D. Moens, "Multivariate dependent interval finite element analysis *via* convex hull pair constructions and the Extended Transformation Method", *Comput. Methods Appl. Mech. Eng.,* vol. 347, pp. 85-102, 2019.
http://dx.doi.org/10.1016/j.cma.2018.12.021

[17] F. Yamazaki, M. Shinozuka, and G. Dasgupta, "Neumann expansion for stochastic finite element analysis", *ASCE J. Engng. Mech.,* vol. 114, pp. 1335-1354, 1988.
http://dx.doi.org/10.1061/(ASCE)0733-9399(1988)114:8(1335)

[18] W. Mo, *Reliability calculations with the stochastic finite element.,* Bentham Science Publishers: Singapore, 2020.
http://dx.doi.org/10.2174/97898114855341200101

[19] S. Srivastava, and D.K. Gupta, "An iterative method for solving general restricted linear equations", *Appl. Math. Comput.,* vol. 262, pp. 344-353, 2015.
http://dx.doi.org/10.1016/j.amc.2015.04.047

[20] S. Dey, T. Mukhopadhyay, and S. Adhikari, *Uncertainty quantification in laminated composites: A meta-model based approach.,* CRC Press, 2018.
http://dx.doi.org/10.1201/9781315155593

Interval Finite Element for Linear Vibration

Abstract: Interval variables have an effect on linear vibration. The linear vibration is transformed into a static problem by Newmark method. The perturbation method, Neumann expansion method, Taylor expansion method, Sherman Morrison Woodbury expansion method and a new iterative method of interval finite element for linear vibration are proposed. The detailed derivation processes are explored.

Keywords: Linear vibration, Perturbation, Neumann expansion, Taylor expansion, Sherman Morrison Woodbury expansion, A new iterative method.

INTRODUCTION

Stochastic finite element has been studied for more than 50 years. Stochastic finite element requires statistical data. Obtaining statistics data is troublesome. Interval finite element does not need probability density function and statistical data. It is difficult to determine the probability density function in engineering applications. Linear vibration is sometimes greatly affected by interval variables.

A combinatorial approach and an inequality-based method are used to solve interval equations [1]. The scatter of external loads identified by displacement input with uncertainty is estimated by the Lagrange multiplier method [2]. An iterative algorithm is a conservative solution for linear interval finite element analysis [3]. The newmethod is based on an element-by-element technique to solveuncertainty in mechanics problems [4]. Anti-optimisation of interval finite element is proposed for uncertain structures analysis [5]. Interval boundary element methods have been explored in finite element analysis with parametric uncertainties [6]. The objective is to give a general overview of non-probabilistic finite element analysis with parametric uncertainty [7]. A method to calculate the static structures with uncertain-but-bounded axial stiffness is proposed [8].The elastic modulus of one-dimensional heterogeneous solids is considered both a probabilistic and a non-probabilistic approach [9]. The present paper is to determine bounds for the stationary stochastic response of truss structures *via* interval analysis [10] .A novel expression of the frequency response function (FRF) matrix of discretized

Wenhui Mo

structures with uncertain-but-bounded parameters are formulated [11]. Reliability analysis of randomly excited structures with interval uncertainties is addressed and the improved interval analysis handles uncertain-but-bounded parameters [12]. A novel Interval finite element method is formulated to improve interval analysis [13]. Finite element analysis of structures is addressed for interval and stochastic finite element analysis [14]. The paper presents the formulation of a stochastic B-spline wavelet on the interval finite element in elasto-statics analysis [15]. The paper presents an interval finite element method based on stochastic B-spline wavelet for beams [16].

Considering the influence of interval variables on linear vibration, five calculation methods of interval finite element for linear vibration are proposed. They are the perturbation method, Neumannexpansion method, Taylor expansion method, Sherman Morrison Woodbury expansion method and a new iterative method. The calculation formulas are given respectively.

Interval Perturbation Finite Element for Linear Vibration

For a linear system, the dynamic equilibrium equation is given by

$$[M]\{\ddot{\delta}\} + [C]\{\dot{\delta}\} + [K]\{\delta\} = \{F\} \tag{1}$$

Where $\{\ddot{\delta}\}, \{\dot{\delta}\}, \{\delta\}$ are the acceleration, velocity and displacement vectors. $[M][K]$ and $[C]$ are the global mass, stiffness and damping matrices obtained by assembling the element variables in the global coordinate system.

For ease of programming, the comprehensive calculation steps of the Newmark method are as follows

The matrices $[K]$, $[M]$ and $[C]$ are formed.

The initial values $\{\delta_t\}, \{\dot{\delta}_t\}, \{\ddot{\delta}_t\}$ are given.

After selecting step Δt and parameters γ, β, the following relevant parameters are calculated:

$$\gamma \geq 0.50 \tag{2}$$

$$\beta \geq 0.25(0.5 + \gamma)^2 \tag{3}$$

$$b_0 = \frac{1}{\beta(\Delta t)^2} \tag{4}$$

$$b_1 = \frac{\gamma}{\beta \Delta t} \tag{5}$$

$$b_2 = \frac{1}{\beta \Delta t} \tag{6}$$

$$b_3 = \frac{1}{2\beta} - 1 \tag{7}$$

$$b_4 = \frac{\gamma}{\beta} - 1 \tag{8}$$

$$b_5 = \frac{\Delta t}{2}\left(\frac{\gamma}{\beta} - 2\right) \tag{9}$$

$$b_6 = \Delta t(1 - \gamma) \tag{10}$$

$$b_7 = \gamma \Delta t \tag{11}$$

The stiffness matrix is defined as

$$[\tilde{K}] = [K] + b_0[M] + b_1[C] \tag{12}$$

The stiffness matrix inversion $\left[\tilde{K}\right]^{-1}$ is solved.

At the time $t + \Delta t$, the load vector is defined as

$$\{\tilde{F}_{t+\Delta t}\} = \{F_{t+\Delta t}\} + [M](b_0\{\delta_t\} +$$

$$b_2\{\dot{\delta}_t\} + b_3\{\ddot{\delta}_t\}) \tag{13}$$

$$+ [C](b_1\{\delta_t\} + b_4\{\dot{\delta}_t\} + b_5\{\ddot{\delta}_t\})$$ At time $t + \Delta t$, the displacement vector is given by

$$\{\delta_{t+\Delta t}\} = [\tilde{K}]^{-1}\{\tilde{F}_{t+\Delta t}\} \tag{14}$$

At the time $t + \Delta t$, the acceleration vector and velocity vector are obtained as

$$\{\ddot{\delta}_{t+\Delta t}\} = b_0(\{\delta_{t+\Delta t}\} - \{\delta_t\}) - b_2\{\dot{\delta}_t\} - b_3\{\ddot{\delta}_t\} \tag{15}$$

$$\{\dot{\delta}_{t+\Delta t}\} = \{\dot{\delta}_t\} + b_6\{\ddot{\delta}_t\} + b_7\{\ddot{\delta}_{t+\Delta t}\} \tag{16}$$

Vectors $\{\delta_{t+i_1\Delta t}\}, \{\dot{\delta}_{t+i_1\Delta t}\}, \{\ddot{\delta}_{t+i_1\Delta t}\}$ are solved at the time $t + i_1\Delta t (i_1 = 2, 3, \cdots, n_1)$ step-by-step.

Material properties, geometry parameters and applied loads are assumed to be interval variables. They are expressed as

$$\bar{a} = a + \Delta a \tag{17}$$

$$\underline{a} = a - \Delta a \tag{18}$$

Where a is midpoints of interval variables, Δa is a small perturbation.

Eq.14 is rewritten

$$[\tilde{K}]\{\delta_{t+\Delta t}\} = \{\tilde{F}_{t+\Delta t}\} \tag{19}$$

Using perturbation technology, we obtain

$$\left(\left[\tilde{K}\right]+\Delta\left[\tilde{K}\right]\right)\left(\left\{\delta_{t+\Delta t}\right\}+\Delta\left\{\delta_{t+\Delta t}\right\}\right)=\left(\left\{\tilde{F}_{t+\Delta t}\right\}+\Delta\left\{\tilde{F}_{t+\Delta t}\right\}\right) \tag{20}$$

where $\Delta\left[\tilde{K}\right]$ is a small perturbation of $\left[\tilde{K}\right], \Delta\{\tilde{F}_{t+\Delta t}\}$ is a small perturbation $\{\tilde{F}_{t+\Delta t}\}$. $\Delta\{\delta_{t+\Delta t}\}$ is a small perturbation of $\{\delta_{t+\Delta t}\}$ in the following equations.

Eq.20 is rewritten

$$\left(\left\{\delta_{t+\Delta t}\right\}+\Delta\left\{\delta_{t+\Delta t}\right\}\right)=\left(\left[\tilde{K}\right]+\Delta\left[\tilde{K}\right]\right)^{-1}\left(\left\{\tilde{F}_{t+\Delta t}\right\}+\Delta\left\{\tilde{F}_{t+\Delta t}\right\}\right) \tag{21}$$

The Neumann expansion of $\left[\tilde{K}\right]+\Delta\left[\tilde{K}\right]^{-1}$ takes the following form

$$\left(\left[\tilde{K}\right]+\Delta\left[\tilde{K}\right]\right)^{-1}=\sum_{i=0}^{\infty}\left(-\left[\tilde{K}\right]^{-1}\Delta\left[\tilde{K}\right]\right)^{i}\left[\tilde{K}\right]^{-1} \tag{22}$$

It is well known that the Neumann expansion converges if the absolute values of all the eigenvalues of $\left[\tilde{K}\right]^{-1}\Delta\left[\tilde{K}\right]$ are less than 1.

Neglecting second-order terms, we get

$$\Delta\left\{\delta_{t+\Delta t}\right\}=\left[\tilde{K}\right]\Delta\left\{\tilde{F}_{t+\Delta t}\right\}-\left[\tilde{K}\right]^{-1}\Delta\left[\tilde{K}\right]\left\{\delta_{t+\Delta t}\right\} \tag{23}$$

$\Delta\left\{\tilde{F}_{t+\Delta t}\right\}$ is given by

$$\Delta\left\{\tilde{F}_{t+\Delta t}\right\}=\Delta\left\{F_{t+\Delta t}\right\}+\Delta\left[M\right]\left(b_{0}\left\{\delta_{t}\right\}+\right.$$

$$\left.b_{2}\left\{\dot{\delta}_{t}\right\}+b_{3}\left\{\ddot{\delta}_{t}\right\}\right)+\left[M\right]\left(b_{0}\Delta\left\{\delta_{t}\right\}+b_{2}\Delta\left\{\dot{\delta}_{t}\right\}+b_{3}\Delta\left\{\ddot{\delta}_{t}\right\}\right)$$

$$+\Delta\left[C\right]\left(b_{1}\left\{\delta_{t}\right\}+b_{4}\left\{\dot{\delta}_{t}\right\}+b_{5}\left\{\ddot{\delta}_{t}\right\}\right)_{+}$$

$$\left[C\right]\left(b_{1}\Delta\left\{\delta_{t}\right\}+b_{4}\Delta\left\{\dot{\delta}_{t}\right\}+b_{5}\Delta\left\{\ddot{\delta}_{t}\right\}\right) \tag{24}$$

obtained.

$\Delta \ \ddot{\delta}_{t+\Delta t}$ is given by

$$\Delta \ \ddot{\delta}_{t+\Delta t} = b_0 \ \Delta \ \delta_{t+\Delta t} \ -\Delta \ \delta_t \ -b_2 \Delta \ \dot{\delta}_t \ -b_3 \Delta \ \ddot{\delta}_t \tag{25}$$

$\Delta \ \dot{\delta}_{t+\Delta t}$ is given by $\Delta \ \dot{\delta}_{t+\Delta t} = \Delta \ \dot{\delta}_t \ +b_6 \Delta \ \ddot{\delta}_t \ +b_7 \Delta \ \ddot{\delta}_{t+\Delta t}$ $\tag{26}$

$\delta_{t+\Delta t}{}'$ can be written as

$$\delta_{t+\Delta t}{}' = \left[\{\underline{\delta}_{t+\Delta t}\}, \{\bar{\delta}_{t+\Delta t}\} \right] = \ \delta_{t+\Delta t}{}^c +\Delta \ \delta_{t+\Delta t}{}'$$
$$= \ \delta_{t+\Delta t}{}^c +\left[-\Delta \ \delta_{t+\Delta t} \ , \Delta \ \delta_{t+\Delta t} \right] \tag{27}$$

We obtain

$$\underline{\delta}_{t+\Delta t} = \{\delta_{t+\Delta t}\}^c -\Delta\{\delta_{t+\Delta t}\} \tag{28}$$

$$\bar{\delta}_{t+\Delta t} = \{\delta_{t+\Delta t}\}^c +\Delta\{\delta_{t+\Delta t}\} \tag{29}$$

where $\delta_{t+\Delta t}{}^c = \left[\tilde{K} \right]^{-1} \ \tilde{F}_{t+\Delta t}$, $\underline{\delta}_{t+\Delta t}$ is lower bound of $\delta_{t+\Delta t}$.

$\bar{\delta}_{t+\Delta t}$ is upper bound of $\delta_{t+\Delta t}$.

Similarly, the lower bound and upper bound of a vector $\{\delta_{t+i_1\Delta t}\}$ can be solved at the time $t+i_1\Delta t \left(i_1 = 2,3,\cdots,n_1 \right)$ step by step.

At the time $t' = t+i_2\Delta t \left(i_2 = 1,2,\cdots,n_1 \right)$, the stress for the element d is given by

$$\{\sigma\} = [D][B]\{\delta_{t'}{}^d\} \tag{30}$$

where, $[D]$ = the material response matrix of the element d, $[B]$= the gradient matrix of element d and $\{\delta_{t'}^d\}$= the element d nodal displacement vector at a time t'. $\Delta\{\sigma\}$ is given by

$$\Delta\{\sigma\} = [D][B]\Delta\{\delta_{t'}^d\} + [D]\Delta[B]\{\delta_{t'}^d\} + \Delta[D][B]\{\delta_{t'}^d\} \tag{31}$$

Where $\Delta\{\sigma\}$ is the perturbation of $\{\sigma\}$

$$\{\bar{\sigma}\} = \{\sigma\} + \Delta\{\sigma\} \tag{32}$$

$$\{\underline{\sigma}\} = \{\sigma\} - \Delta\{\sigma\} \tag{33}$$

where $\{\bar{\sigma}\}$ is upper bound of $\{\sigma\}$, $\{\underline{\sigma}\}$ is lower bound of $\{\sigma\}$.

Interval Neumann finite Element for Linear Vibration

Neumann expansion is applied to stochastic finite elements[17][18]. Neumann expansion is also applied to interval finite element for linear vibration. Interval variable $[\underline{a}, \bar{a}]$ is generated by the following formula

$$a_i = \underline{a} + \frac{\bar{a} - \underline{a}}{n} i - \frac{\bar{a} - \underline{a}}{2} = \frac{(3n - 2i)\underline{a} + (2i - n)\bar{a}}{2n} \tag{34}$$

$$i = 1, 2, \cdots, n$$

Material properties, geometry parameters and applied loads of structures are assumed to be interval variables.

At the time $t + \Delta t$, the displacement vector using Newmark method is given by

$$\{\delta_{t+\Delta t}\} = [\tilde{K}]^{-1} \{\tilde{F}_{t+\Delta t}\} \tag{35}$$

Neumann expansion of $[\tilde{K}]^{-1}$ takes the following form:

$$[\tilde{K}]^{-1} = ([\tilde{K}_0] + \Delta[\tilde{K}])^{-1} = (I - P + P^2 - P^3 + \cdots)[\tilde{K}_0]^{-1} \tag{36}$$

$\{\delta_{t+\Delta t}\}$ is represented by the following series as

$$\{\delta_{t+\Delta t}\} = \{\delta_{t+\Delta t_0}\} - \{\delta_{t+\Delta t_1}\} + \{\delta_{t+\Delta t_2}\} - \{\delta_{t+\Delta t_3}\} + \cdots \tag{37}$$

This series solution is equivalent to the following equation:

$$[K_0]\{\delta_{t+\Delta t_i}\} = \Delta[K]_{i-1}\{\delta_{t+\Delta t_{i-1}}\} \quad i=1,2, \cdots \tag{38}$$

Substituting N_1 samples of interval variables into above equations, the vectors N_1 $\{\delta_{t+\Delta t}\}$ can be obtained.

The maximums and minimums of N_1 $\{\delta_{t+\Delta t}\}$ are the lower bounds and upper bounds of $\{\delta_{t+\Delta t}\}$.

$$Upper_x = \max(\{\delta_{t+\Delta t}\}_{1x}, \{\delta_{t+\Delta t}\}_{2x}, \cdots, \{\delta_{t+\Delta t}\}_{N_1 x}) \tag{39}$$

where max( ) and $Upper_x$ see chapter 4.

$$Lower_x = \min(\{\delta_{t+\Delta t}\}_{1x}, \{\delta_{t+\Delta t}\}_{2x}, \cdots, \{\delta_{t+\Delta t}\}_{N_1 x}) \tag{40}$$

where min( ) and $Lower_x$ see chapter 4.

$$Upper_y = \max(\{\delta_{t+\Delta t}\}_{1y}, \{\delta_{t+\Delta t}\}_{2y}, \cdots, \{\delta_{t+\Delta t}\}_{N_1 y}) \tag{41}$$

where max( ) and $Upper_y$ see chapter 4

$$Lower_y = \min(\{\delta_{t+\Delta t}\}_{1y}, \{\delta_{t+\Delta t}\}_{2y}, \cdots, \{\delta_{t+\Delta t}\}_{N_1 y}) \tag{42}$$

where min( ) and $Lower_y$ see chapter 4.

$$Upper_z = \max(\{\delta_{t+\Delta t}\}_{1z}, \{\delta_{t+\Delta t}\}_{2z}, \cdots, \{\delta_{t+\Delta t}\}_{N_1 z}) \tag{43}$$

where max ( ) is and *Upper*$_z$ see chapter 4.

$$Lower_z = \min(\{\delta_{t+\Delta t}\}_{1z}, \{\delta_{t+\Delta t}\}_{2z}, \cdots, \{\delta_{t+\Delta t}\}_{N_1 z}) \tag{44}$$

where min( ) and *Lower*$_z$ see chapter 4.

At the time $t + \Delta t$, the velocity vector and acceleration vector are obtained as

$$\{\ddot{\delta}_{t+\Delta t}\} = b_0 (\{\delta_{t+\Delta t}\} - \{\delta_t\}) - b_2 \{\dot{\delta}_t\} - b_3 \{\ddot{\delta}_t\} \tag{45}$$

$$\{\dot{\delta}_{t+\Delta t}\} = \{\dot{\delta}_t\} + b_6 \{\ddot{\delta}_t\} + b_7 \{\ddot{\delta}_{t+\Delta t}\} \tag{46}$$

Then, the upper bounds and lower bounds of displacement are obtained at a time $t + i_1 \Delta t (i_1 = 2, 3, \cdots, n_1)$ step by step.

At the time $t' = t + i_2 \Delta t (i_2 = 1, 2, \cdots, n_1)$, the stress for the element d is given by

$$\{\sigma\} = [D][B]\{\delta_{t'}^d\} \tag{47}$$

where, $[D]$ =the material response matrix of the element d, $[B]$ = the gradient matrix of element d and $\{\delta_{t'}^d\}$ =the element d nodal displacement vector at a time t'.

Substituting N_1 samples of random variables $a_1, a_2, \cdots, a_i, \cdots, a_{n_1}$ and $N_1 \{\delta_{t'}^d\}$ into Eq.47, the vectors $\{\sigma\}_1, \{\sigma\}_2, \cdots, \{\sigma\}_{N_1}$ can be obtained. The maximums and minimums of $\{\sigma\}_1, \{\sigma\}_2, \cdots, \{\sigma\}_{N_1}$ are the lower bounds and upper bounds of $\{\sigma\}$.

$$Upper_{\sigma x} = \max(\sigma_{1x}, \sigma_{2x}, \cdots, \sigma_{N_1 x}) \tag{48}$$

where max( ) and *Upper*$_x$ see chapter 4.

$$Lower_{\sigma x} = \min(\sigma_{1x}, \sigma_{2x}, \cdots, \sigma_{N_1 x}) \tag{49}$$

where min( ) and $Lower_{\sigma x}$ see chapter 4.

$$Upper_{\sigma y} = \max(\sigma_{1y}, \sigma_{2y}, \cdots, \sigma_{N_1 y}) \tag{50}$$

where max( ) and $Upper_{\sigma y}$ see chapter 4

$$Lower_{\sigma y} = \min(\sigma_{1y}, \sigma_{2y}, \cdots, \sigma_{N_1 y}) \tag{51}$$

where min( ) and $Lower_{\sigma y}$ see chapter 4.

$$Upper_{\sigma z} = \max(\sigma_{1z}, \sigma_{2z}, \cdots, \sigma_{N_1 z}) \tag{52}$$

where max( ) and $Upper_{\sigma z}$ see chapter 4.

$$Lower_{\sigma z} = \min(\sigma_{1z}, \sigma_{2z}, \cdots, \sigma_{N_1 z}) \tag{53}$$

where min( ) and $Lower_{\sigma z}$ see chapter 4.

$$Upper_{\tau xy} = \max(\tau_{1xy}, \tau_{2xy}, \cdots, \tau_{N_1 xy}) \tag{54}$$

where max( ) and $Upper_{\tau xy}$ see chapter 4.

$$Lower_{\tau xy} = \min(\tau_{1xy}, \tau_{2xy}, \cdots, \tau_{N_1 xy}) \tag{55}$$

where min( ) and $Lower_{\tau xy}$ see chapter 4.

$$Upper_{\tau yz} = \max(\tau_{1yz}, \tau_{2yz}, \cdots, \tau_{N_1 yz}) \tag{56}$$

where max( ) and $Upper_{\tau yz}$ see chapter 4.

$$Lower_{\tau yz} = \min(\tau_{1yz}, \tau_{2yz}, \cdots, \tau_{N_1 yz}) \tag{57}$$

where min( ) and $Lower_{\tau yz}$ see chapter 4.

$$Upper_{\tau zx} = \max(\tau_{1zx}, \tau_{2zx}, \cdots, \tau_{N_1 zx}) \tag{58}$$

where max( ) and $Upper_{\tau yz}$ see chapter 4.

$$Lower_{\tau zx} = \min(\tau_{1zx}, \tau_{2zx}, \cdots, \tau_{N_1 zx}) \tag{59}$$

where min( ) and $Lower_{\tau zx}$ see chapter 4.

Interval Taylor Finite Element for Linear Vibration

Interval variable $\left[\underline{a}, \overline{a}\right]$ is generated by the following formula

$$a_i = \underline{a} + \frac{\overline{a} - \underline{a}}{n} i = \frac{i\overline{a} + (n-i)\underline{a}}{n} \tag{60}$$

$$i = 1, 2, \cdots, n$$

Material properties, geometry parameters and applied loads of structures are assumed to be interval variables. They are regarded as N interval variables $[\underline{a}_1, \overline{a}_1], [\underline{a}_2, \overline{a}_2], \cdots, [\underline{a}_j, \overline{a}_j], \cdots, [\underline{a}_n, \overline{a}_n]$.

At the time $t + \Delta t$, the displacement vector using Newmark method is given by

$$\{\delta_{t+\Delta t}\} = \left[\tilde{K}\right]^{-1} \{\tilde{F}_{t+\Delta t}\} \tag{61}$$

Eq.61 can be rewritten

$$\left[\tilde{K}\right]\{\delta_{t+\Delta t}\} = \{\tilde{F}_{t+\Delta t}\} \tag{62}$$

The partial derivative of Eq.62 with respect to a_i is given by

$$\frac{\partial \{\delta_{t+\Delta t}\}}{\partial a_i} = \left[\tilde{K}\right]^{-1} \left(\frac{\partial \{\tilde{F}_{t+\Delta t}\}}{\partial a_i} - \frac{\partial \left[\tilde{K}\right]}{\partial a_i} \{\delta_{t+\Delta t}\} \right) \tag{63}$$

where see Chapter 2 for the formula of $\dfrac{\partial \left\{ \tilde{F}_{t+\Delta t} \right\}}{\partial a_i}$.

After $\dfrac{\partial \{\delta_t\}}{\partial a_i} = q_0, \dfrac{\partial \{\dot{\delta}_t\}}{\partial a_i} = \dot{q}_0, \dfrac{\partial \{\ddot{\delta}_t\}}{\partial a_i} = \ddot{q}_0$ are given, $\dfrac{\partial \left\{ \tilde{F}_{t+\Delta t} \right\}}{\partial a_i}$ can be calculated.

The partial derivative of Eq.63 with respect to a_j is given by

$$\frac{\partial^2 \left\{ \delta_{t+\Delta t} \right\}}{\partial a_i \partial a_j} = \left[\tilde{K} \right]^{-1} \left(\frac{\partial^2 \left\{ \tilde{F}_{t+\Delta t} \right\}}{\partial a_i \partial a_j} - \frac{\partial \left[\tilde{K} \right]}{\partial a_i} \frac{\partial \left\{ \delta_{t+\Delta t} \right\}}{\partial a_j} \right.$$

$$\left. - \frac{\partial \left[\tilde{K} \right]}{\partial a_j} \frac{\partial \left\{ \delta_{t+\Delta t} \right\}}{\partial a_i} - \frac{\partial^2 \left[\tilde{K} \right]}{\partial a_i \partial a_j} \{ \delta_{t+\Delta t} \} \right) \tag{64}$$

where see Chapter 2 for the formula of $\dfrac{\partial^2 \left\{ \tilde{F}_{t+\Delta t} \right\}}{\partial a_i \partial a_j}$

After $\dfrac{\partial \{\delta_t\}}{\partial a_j} = q_1, \dfrac{\partial \{\delta_t\}}{\partial a_j} = q_1, \dfrac{\partial \{\dot{\delta}_t\}}{\partial a_j} = \dot{q}_1, \dfrac{\partial \{\ddot{\delta}_t\}}{\partial a_j} = \ddot{q}_1 , \dfrac{\partial^2 \{\delta_t\}}{\partial a_i \partial a_j} = r_0, \dfrac{\partial^2 \{\dot{\delta}_t\}}{\partial a_i \partial a_j} = \dot{r}_0,$

$\dfrac{\partial^2 \{\ddot{\delta}_t\}}{\partial a_i \partial a_j} = \ddot{r}_0$ are given, $\dfrac{\partial^2 \left\{ \tilde{F}_{t+\Delta t} \right\}}{\partial a_i \partial a_j}$ can be calculated.

The displacement is expanded at the midpoints of the interval variables by means of a Taylor series.

The second-order term of the Taylor expansion formula is given by

$$\{\delta_{t+\Delta t}\} \approx \{\delta_{t+\Delta t}\}^0 + \sum_{k=1}^{2} \frac{1}{k!} \sum_{i_1,i_2,\cdots,i_k=1}^{n} \frac{\partial^k \{\delta_{t+\Delta t}\}}{\partial a_{i_1} \partial a_{i_2} \cdots \partial a_{i_k}} (a^0)(a_{i_1} - a^0_{i_1})$$

$$\tag{65}$$

$$(a_{i_2} - a^0_{i_2}) \cdots (a_{i_k} - a^0_{i_k})$$

Substituting N_1 samples of interval variables into Eq.65, the vector $N_1 \{\delta_{t+\Delta t}\}$ can be obtained.

The maximums and minimums of component of $N_1 \{\delta_{t+\Delta t}\}$ are the upper bounds and lower bounds of $\{\delta_{t+\Delta t}\}$.

The partial derivative of $\ddot{\delta}_{t+\Delta t}$ with respect to a_i is given by

$$\frac{\partial \{\ddot{\delta}_{t+\Delta t}\}}{\partial a_i} = b_0 \left(\frac{\partial \{\delta_{t+\Delta t}\}}{\partial a_i} - \frac{\partial \{\delta_t\}}{\partial a_i} \right)$$

$$-b_2 \frac{\partial \{\dot{\delta}_t\}}{\partial a_i} - b_3 \frac{\partial \{\ddot{\delta}_t\}}{\partial a_i} \tag{66}$$

The partial derivative of $\dot{\delta}_{t+\Delta t}$ with respect to a_i is given by

$$\frac{\partial \{\dot{\delta}_{t+\Delta t}\}}{\partial a_i} = \frac{\partial \{\dot{\delta}_t\}}{\partial a_i} + b_6 \frac{\partial \{\ddot{\delta}_t\}}{\partial a_i} + b_7 \frac{\partial \{\ddot{\delta}_{t+\Delta t}\}}{\partial a_i} \tag{67}$$

The partial derivative of Eq.66 with respect to a_j is given by

$$\frac{\partial^2 \{\ddot{\delta}_{t+\Delta t}\}}{\partial a_i \partial a_j} = b_0 \left(\frac{\partial^2 \{\delta_{t+\Delta t}\}}{\partial a_i \partial a_j} - \frac{\partial^2 \{\delta_t\}}{\partial a_i \partial a_j} \right) - b_2 \frac{\partial^2 \{\dot{\delta}_t\}}{\partial a_i \partial a_j} - b_3 \frac{\partial^2 \{\ddot{\delta}_t\}}{\partial a_i \partial a_j} \tag{68}$$

The partial derivative of Eq.67 with respect to a_j is given by

$$\frac{\partial^2 \{\dot{\delta}_{t+\Delta t}\}}{\partial a_i \partial a_j} = \frac{\partial^2 \{\dot{\delta}_t\}}{\partial a_i \partial a_j} + b_6 \frac{\partial^2 \{\ddot{\delta}_t\}}{\partial a_i \partial a_j} + b_7 \frac{\partial^2 \{\ddot{\delta}_{t+\Delta t}\}}{\partial a_i \partial a_j} \tag{69}$$

Eqs 66, 67, 68 and 69 must be calculated for the following iteration.

Then, the upper bounds and lower bounds of displacement are obtained at a time $t + i_1 \Delta t \, (i_1 = 2, 3, \cdots, n_1)$ step by step. See chapter 4 for the calculation formula of upper and lower bounds of displacement.

At time $t' = t + i_2 \Delta t \, (i_2 = 1, 2, \cdots, n_1)$, the stress for the element d is given by

$$\{\sigma\} = [D][B]\{\delta_{t'}^d\} \tag{70}$$

Substituting N_1 samples of random variables $a_1, a_2, \cdots, a_i, \cdots, a_{n_1}$ and $N_1 \{\delta_{t'}^d\}$ into Eq.70, the vectors $\{\sigma\}_1, \{\sigma\}_2, \cdots, \{\sigma\}_{N_1}$ can be obtained.

The maximums and minimums of $\{\sigma\}_1, \{\sigma\}_2, \cdots, \{\sigma\}_{N_1}$ are the lower bounds and upper bounds of $\{\sigma\}$. See chapter 4 for the calculation formula of upper and lower bounds of stress.

Interval Sherman-Morrison-Woodbury Expansion Finite Element for Linear Vibration

Interval   variable $[\underline{a}, \overline{a}]$ is   generated   by   the   following   formula

$$a_i = \underline{a} + \frac{\overline{a} - \underline{a}}{n} i - \frac{\overline{a} - \underline{a}}{2} = \frac{(3n - 2i)\underline{a} + (2i - n)\overline{a}}{2n} \tag{71}$$

$i = 1, 2, \cdots, n$

Material properties, geometry parameters and applied loads of structures are assumed to be interval variables.

At the time $t + \Delta t$, the displacement vector using Newmark method is given by

$$\{\delta_{t+\Delta t}\} = \left[\tilde{K}\right]^{-1} \{\tilde{F}_{t+\Delta t}\} \tag{72}$$

Eq72. is rewritten as

$$U = K^{-1}F \tag{73}$$

Sherman-Morrison-Woodbury expansion is given by

$$(K + UV^T)^{-1} = K^{-1} - K^{-1}U(I + V^T K^{-1}U)^{-1}V^T K^{-1} \tag{74}$$

We obtain

$$(K_0 + (\Delta K^T)^T)^{-1} = K_0^{-1} - K_0^{-1}\left(I + (\Delta K^T)^T K_0^{-1}\right)^{-1}(\Delta K^T)^T K_0^{-1} \tag{75}$$

$$U = (K_0 + (\Delta K^T)^T)^{-1}F = K_0^{-1}F - K_0^{-1}\left(I + (\Delta K^T)^T K_0^{-1}\right)^{-1}(\Delta K^T)^T K_0^{-1}F \tag{76}$$

Using Neumann expansion, the above formula is

$$U = (K_0 + (\Delta K^T)^T)^{-1}F = K_0^{-1}F - K_0^{-1}\left(-(\Delta K^T)^T K_0^{-1} + \left((\Delta K^T)^T K_0^{-1}\right)^2\right)$$

$$(\Delta K^T)^T K_0^{-1}F \tag{77}$$

Substituting N_1 samples of interval variables into Eq.76 and Eq.77, the vectors $U_1, U_2, \cdots, U_{N_1}$ can be obtained.

The maximums and minimums of $U_1, U_2, \cdots, U_{N_1}$ are the lower bounds and upper bounds of $U_1, U_2, \cdots, U_{N_1}$.

At the $t + \Delta t$, the velocity vector and acceleration vector are obtained as

$$\{\ddot{\delta}_{t+\Delta t}\} = b_0(\{\delta_{t+\Delta t}\} - \{\delta_t\}) - b_2\{\dot{\delta}_t\} - b_3\{\ddot{\delta}_t\} \tag{78}$$

$$\{\dot{\delta}_{t+\Delta t}\} = \{\dot{\delta}_t\} + b_6\{\ddot{\delta}_t\} + b_7\{\ddot{\delta}_{t+\Delta t}\} \tag{79}$$

Vectors $\{\delta_{t+i_1\Delta t}\}, \{\dot{\delta}_{t+i_1\Delta t}\}, \{\ddot{\delta}_{t+i_1\Delta t}\}$ are solved at time $t + i_1\Delta t$ $(i_1 = 2,3,\cdots,n_1)$ step-by-step. See chapter 4 for the calculation formula of upper and lower bounds of displacement.

The stress for the element d is given by

$$\{\sigma\} = [D][B]\{\delta_{t'}^d\} \tag{80}$$

Substituting N_1 samples of random variables $a_1, a_2, \cdots, a_i, \cdots, a_{n_1}$ and $N_1 \{\delta_{t'}^d\}$ into Eq.80, the vectors $\{\sigma\}_1, \{\sigma\}_2, \cdots, \{\sigma\}_{N_1}$ can be obtained.

The maximums and minimums of $\{\sigma\}_1, \{\sigma\}_2, \cdots, \{\sigma\}_{N_1}$ are the lower bounds and upper bounds of $\{\sigma\}_1, \{\sigma\}_2, \cdots, \{\sigma\}_{N_1}$. See chapter 4 for the calculation formula of upper and lower bounds of stress.

A New Iterative Method (NIM)

Interval variable $\left[\underline{a}, \overline{a}\right]$ is generated by the following formula

$$a_i = \underline{a} + \frac{\overline{a} - \underline{a}}{n} i = \frac{i\overline{a} + (n-i)\underline{a}}{n} \tag{81}$$

$$i = 1, 2, \cdots, n$$

Material properties, geometry parameters and applied loads of structures are assumed to be interval variables.

At the time $t + \Delta t$, the displacement vector using Newmark method is given by

$$\{\delta_{t+\Delta t}\} = \left[\tilde{K}\right]^{-1} \{\tilde{F}_{t+\Delta t}\} \tag{82}$$

Eq. 82 is rewritten as

Ax=b, $\tag{83}$

In this section, a new iterative method (NIM)and its convergence analysis for finding a solution of Eq.83 ,along with the estimation of error bounds are described. The iterative method [19]is defined for k=0,1,2,... by

$$Z_{k+1}=Z_k(2I-AZ_k) \tag{84}$$

$$x_{k+1}=x_k+Z_{k+1}(b-Ax_k) \tag{85}$$

Substituting N_1 samples of interval variables into Eq.84 and Eq.85, the vectors N_1 $\{\delta_{t+\Delta t}\}$ can be obtained.

The maximums and minimums of N_1 $\{\delta_{t+\Delta t}\}$ are the lower bounds and upper bounds of $\{\delta_{t+\Delta t}\}$.

At the time $t+\Delta t$, the velocity vector and acceleration vector are obtained as

$$\{\ddot{\delta}_{t+\Delta t}\} = b_0\left(\{\delta_{t+\Delta t}\}-\{\delta_t\}\right)-b_2\{\dot{\delta}_t\}-b_3\{\ddot{\delta}_t\} \tag{86}$$

$$\{\dot{\delta}_{t+\Delta t}\} = \{\dot{\delta}_t\}+b_6\{\ddot{\delta}_t\}+b_7\{\ddot{\delta}_{t+\Delta t}\} \tag{87}$$

Vectors $\{\delta_{t+i_1\Delta t}\},\{\dot{\delta}_{t+i_1\Delta t}\},\{\ddot{\delta}_{t+i_1\Delta t}\}$ are solved at time $t+i_1\Delta t\left(i_1=2,3,\cdots,n_1\right)$ step-by-step. See chapter 4 for the calculation formula of upper and lower bounds of displacement.

At the time $t'=t+i_2\Delta t\left(i_2=1,2,\cdots,n_1\right)$, the stress for element d is given by

$$\{\sigma\}=[D][B]\{\delta_{t'}^d\} \tag{88}$$

Substituting N_1 samples of random variables $a_1,a_2,\cdots,a_i,\cdots,a_{n_1}$ and $N_1\{\delta_{t'}^d\}$ into Eq.88, the vectors $\{\sigma\}_1,\{\sigma\}_2,\cdots,\{\sigma\}_{N_1}$ can be obtained.

The maximums and minimums of $\{\sigma\}_1, \{\sigma\}_2, \cdots, \{\sigma\}_{N_1}$ are the lower bounds and upper bounds of $\{\sigma\}$. See chapter 4 for the calculation formula of upper and lower bounds of stress.

CONCLUDING REMARKS

The linear differential equations are transformed into linear equations by Newmark method. The linear vibration problem becomes a static problem. Five calculation methods of interval finite element for linear vibration are proposed. These five methods can be used to calculate practical engineering problems.

REFERENCES

[1]   S.S. Rao, and L. Berke, "Analysis of Uncertain Structural Systems Using Interval Analysis", *AIAA J.,* vol. 35, no. 4, pp. 727-735, 1997.
      http://dx.doi.org/10.2514/2.164

[2]   S. Nakagiri, and K. Suzuki, "Finite element interval analysis of external loads identified by displacement input with uncertainty", *Comput. Methods Appl. Mech. Eng.,* vol. 168, no. 1-4, pp. 63-72, 1999.
      http://dx.doi.org/10.1016/S0045-7825(98)00134-0

[3]   O. Dessombz, F. Thouverez, J-P. La^ın'e, and L. J'ez'eque, "Analysis of Mechanical Systems using Interval Computations applied to Finite Elements Methods", *J. Sound Vibrat.,* no. January, pp. 1-21, 2000.

[4]   R.L. Muhanna, and R.L. Mullen, "Uncertainty in mechanics problems-interval-based approach", *J. Eng. Mech.,* vol. 127, no. JUNE, pp. 557-566, 2001.
      http://dx.doi.org/10.1061/(ASCE)0733-9399(2001)127:6(557)

[5]   S. Mc William, "Anti-optimisation of uncertain structures using interval analysis", *Comput. Struc.,* vol. 79, no. May, pp. 421-430, 2001.
      http://dx.doi.org/10.1016/S0045-7949(00)00143-7

[6]   F. Bart, Bart F. Zalewski, Robert L. Mullen, Rafi L. Muhanna, " Interval boundary element method in the presence of uncertain boundary conditions, integration errors, and truncation errors ", *Engineering Analysis with Boundary Elements,* vol. 33, pp. 508-513, 2009.

[7]   D. Moens, and M. Hanss, "Non-probabilistic finite element analysis for parametric uncertainty treatment in applied mechanics: Recent advances", *Finite Elem. Anal. Des.,* vol. 47, no. 1, pp. 4-16, 2011.
      http://dx.doi.org/10.1016/j.finel.2010.07.010

[8]   N. Impollonia, and G. Muscolino, "Interval analysis of structures with uncertain-but-bounded axial stiffness", *Comput. Methods Appl. Mech. Eng.,* vol. 200, no. 21-22, pp. 1945-1962, 2011.
      http://dx.doi.org/10.1016/j.cma.2010.07.019

[9]   G. Muscolino, A. Sofi, and M. Zingales, "One-dimensional heterogeneous solids with uncertain elastic modulus in presence of long-range interactions: interval versus stochastic analysis", *Comput. Struc.,* vol. 122, pp. 217-229, 2013.
      http://dx.doi.org/10.1016/j.compstruc.2013.03.005

[10] G. Muscolino, and A. Sofi, "Bounds for the stationary stochastic response of truss structures with uncertain-but-bounded parameters", *Mech. Syst. Signal Process.,* vol. 37, no. 1-2, pp. 163-181, 2013.

http://dx.doi.org/10.1016/j.ymssp.2012.06.016

[11] G. Muscolino, R. Santoro, and A. Sofi, "Explicit frequency response functions of discretized structures with uncertain parameters", *Comput. Struc.,* vol. 133, pp. 64-78, 2014.

http://dx.doi.org/10.1016/j.compstruc.2013.11.007

[12] G. Muscolino, R. Santoro, and A. Sofi, "Reliability analysis of structures with interval uncertainties under stationary stochastic excitations", *Comput. Methods Appl. Mech. Eng.,* vol. 300, pp. 47-69, 2016.

http://dx.doi.org/10.1016/j.cma.2015.10.023

[13] A. Sofi, and E. Romeo, "A novel Interval Finite Element Method based on the improved interval analysis", *Comput. Methods Appl. Mech. Eng.,* vol. 311, pp. 671-697, 2016.

http://dx.doi.org/10.1016/j.cma.2016.09.009

[14] A. Sofi, and E. Romeo, "A unified response surface framework for the interval and stochastic finite element analysis of structures with uncertain parameters", *Probab. Eng. Mech.,* vol. 54, pp. 25-36, 2018.

http://dx.doi.org/10.1016/j.probengmech.2017.06.004

[15] S. Vadlamani, and C.O. Arun, "A stochastic B-spline wavelet on the interval finite element method for problems in elasto-statics", *Probab. Eng. Mech.,* vol. 58, 2019.102996

http://dx.doi.org/10.1016/j.probengmech.2019.102996

[16] C.O. Shashank Vadlamani, Arun, "A stochastic B-spline wavelet on the interval finite element method for beams", *Comput. Struc.,* vol. 233, 2020.106246

http://dx.doi.org/10.1016/j.compstruc.2020.106246

[17] F. Yamazaki, M. Shinozuka, and G. Dasgupta, "Neumann expansion for stochasticfinite element analysis", *J. Eng. Mech.,* vol. 114, no. 8, pp. 1335-1354, 1988.

http://dx.doi.org/10.1061/(ASCE)0733-9399(1988)114:8(1335)

[18] W. Mo, *Reliability calculations with the stochastic finite element.,* Bentham Science Publishers: Singapore, 2020.

http://dx.doi.org/10.2174/97898114855341200101

[19] S. Srivastava, and D.K. Gupta, "An iterative method for solving general restricted linear equations", *Appl. Math. Comput.,* vol. 262, pp. 344-353, 2015.

http://dx.doi.org/10.1016/j.amc.2015.04.047

[20] S. Dey, T. Mukhopadhyay, and S. Adhikari, *Uncertainty quantification in laminated composites: A meta-model based approach.,* CRC Press, 2018.

http://dx.doi.org/10.1201/9781315155593

CHAPTER 6

Nonlinear Interval Finite Element

Abstract: Nonlinear structures in engineering are affected by uncertain parameters. Firstly, the displacement when the interval variable takes the midpoint value is obtained, and the nonlinear problem is transformed into a linear problem. Five calculation methods of nonlinear interval finite element for general nonlinear problems and elastoplastic problems are proposed. According to the perturbation technique, a perturbation method is proposed. According to Taylor expansion, Taylor expansion method is proposed. Neumann expansion, Sherman Morrison Woodbury expansion and a new iterative method are proposed.

Keywords: Nonlinear structures, Elastoplastic problems, Perturbation method, Taylor expansion method, Neumann expansion, Sherman Morrison Woodbury expansion , A new iterative method.

INTRODUCTION

The influence of uncertain parameters on nonlinear structures can not be ignored sometimes. In order to improve the calculation accuracy of finite element, it is necessary to study the interval finite element of nonlinear structure. The influence of interval variables on engineering problems should be paid attention to. A static structural analysis problem with uncertain parameters can beexpressed a system of linear interval equations [1]. The sensitivity analysis of interval finite element is evaluated [2]. A new formulation has been given for the analysis of mechanical systems using interval finite elements methods [3]. A modified interval perturbation analysis of uncertain structures is presented [4]. Under loading, material and geometric uncertainty, a very sharp enclosure for the solution is obtained [5]. In order to reduce the calculation time, Interval and fuzzy dynamic analysis of finite elements are proposed [6]. The merits of the new approach are demonstrated by computing linear systems with large uncertainties, with applications to truss structures, leading to over 5000 variables and over 10000 interval parameters [7]. This paper presents a novel method to solve interval finite element analysis [8]. A new interval finite element states sharp displacement bounds are produced by the Lagrange multiplier method [9]. Intervals describing variation are parameterized by trigonometric functions [10]. The method is adopted to improve the ordinary interval analysis, based on the so-called affine arithmetic [11]. Chosen numerical algorithms for interval finite element analysis are compared

Wenhui Mo

with the Monte Carlo method [12]. The improved interval analysis *via* extra unitary interval is proposed [13]. The principal idea is to solve an inverse problem for analyzing multivariate interval uncertainty [14]. A response surface approach is adopted for uncertainty propagation analysis which provides a method of interval finite element [15]. One model parameter over the domain is usually modelled using a series expansion for interval finite element analysis via convex hull pair constructions [16].

Five methods for interval finite element calculation of nonlinear structures are proposed. They are perturbation method, Taylor expansion method, Neumann expansion, Sherman MorrisonWoodbury expansion and a new iterative method. The detailed derivation of five calculation methods is given respectively.

General Nonlinear Problems

When the material stress-strain is nonlinear, the stiffness matrix is not constant, which is related to strain and displacement. The global equilibrium equation of the structure is the following nonlinear equations.

$$\{\emptyset\} = K\{\delta\} - \{F\} = 0 \tag{1}$$

where $[K(\delta)]$ is global stiffness matrix, $\{\delta\}$ is displacement matrix and

$\{F\}$ is a load matrix.

Midpoint values of the elastic modulus, Poisson's ratio and load are substituted into the above formula. The displacement is calculated by the tangent stiffness method or initial stress method (See Chapter 1 for details). $\{\delta\}_{n+1}$ is the solution of Eq.1 .After $\{\delta\}_{n+1}$ represents Eq.1, it becomes a linear equation containing interval variables of material properties, geometry parameters and applied loads of structures.

Perturbation technology for nonlinear interval finite element

Material properties, geometry parameters and applied loads are assumed to be interval variables. They are expressed as

$$\bar{a} = a + \Delta a \tag{2}$$

$$\underline{a} = a - \Delta a \tag{3}$$

Where a is midpoints of interval variables, Δa is a small perturbation.

Eq.1 is rewritten

$$[K]\{\delta\} = \{F\} \tag{4}$$

Using perturbation technology and representing $\{\delta\}_{n+1}$, we obtain

$$\left([K] + \Delta[K]\right)\left(\{\delta\} + \Delta\{\delta\}\right) = \left(\{F\} + \Delta\{F\}\right) \tag{5}$$

where $\Delta[K]$ is a small perturbation of $[K]$, $\Delta\{F\}$ is a small perturbationof $\{F\}$. $\Delta\{\delta\}$ is a small perturbation of $\{\delta\}$ in the following equations.

Eq.5 is rewritten

$$\left(\{\delta\} + \Delta\{\delta\}\right) = \left([K] + \Delta[K]\right)^{-1}\left(\{F\} + \Delta\{F\}\right) \tag{6}$$

The Neumann expansion of $\left[\tilde{K}\right] + \Delta\left[\tilde{K}\right]^{-1}$ takes the following form:

$$\left([K] + \Delta[K]\right)^{-1} = \sum_{i=0}^{\infty}\left(-[K]^{-1}\Delta[K]\right)^{i}[K]^{-1} \tag{7}$$

Neglecting second-order terms, we get

$$\Delta\{\delta\} = [K]\Delta\{F\} - [K]^{-1}\Delta[K]\{\delta\} \tag{8}$$

δ^{I} can be written as

$$\delta^{I} = \left[\{\underline{\delta}\}, \{\overline{\delta}\}\right] = \delta^{c} + \Delta\delta^{I} = \delta^{c} + \left[-\Delta\delta, \Delta\delta\right] \tag{9}$$

We obtain

$$\underline{\delta} = \{\delta\}^c - \Delta\{\delta\} \tag{10}$$

$$\overline{\delta} = \{\delta\}^c + \Delta\{\delta\} \tag{11}$$

where $\delta^{\,c} = \left[\tilde{K}\right]^{-1} \tilde{F}$, $\underline{\delta}$ is lower bound of δ ,

$\overline{\delta}$ is upper bound of δ .

Taylor expansion for nonlinear interval finite element

Interval variable   is generated by the following formula

$$a_i = \underline{a} + \frac{\overline{a} - \underline{a}}{n} i = \frac{i\overline{a} + (n-i)\underline{a}}{n} \tag{12}$$

$i = 1, 2, \cdots, n$

Material properties, geometry parameters and applied loads of structures are assumed to be interval variables. They are $[\underline{a}_1, \overline{a}_1], [\underline{a}_2, \overline{a}_2], \cdots, [\underline{a}_j, \overline{a}_j], \cdots, [\underline{a}_n, \overline{a}_n]$.

The equilibrium equation is written as

$$KU = F \tag{13}$$

where U =the displacement vector, F =the external force, K =the global stiffness matrix.

By applying Taylor series at midpoints of interval variables, the following equations are given by

$$U^0 = \left(K^0\right)^{-1} F^0 \tag{14}$$

$$\frac{\partial U}{\partial a_i} = \left(K^0\right)^{-1}\left(\frac{\partial F}{\partial a_i} - \frac{\partial K}{\partial a_i}U^0\right) \tag{15}$$

$$\frac{\partial^2 U}{\partial a_i \partial a_j} = \left(K^0\right)^{-1}\left(\frac{\partial^2 F}{\partial a_i \partial a_j} - \frac{\partial^2 K}{\partial a_i \partial a_j}U^0 - \frac{\partial K}{\partial a_i}\frac{\partial U}{\partial a_j} - \frac{\partial K}{\partial a_j}\frac{\partial U}{\partial a_i}\right) \tag{16}$$

$$\frac{\partial^3 U}{\partial a_i^2 \partial a_j} = (K^0)^{-1}\left(\frac{\partial^3 F}{\partial a_i^2 \partial a_j} - \frac{\partial^3 K}{\partial a_i^2 \partial a_j}U^0 - 3\frac{\partial^2 K}{\partial a_i \partial a_j}\frac{\partial U}{\partial a_i} - 3\frac{\partial K}{a_i}\frac{\partial^2 U}{\partial a_i \partial a_j}\right) \tag{17}$$

The second-order term of the Taylor expansion formula is given by

$$U \approx U^0 + \sum_{k=1}^{2}\frac{1}{k!}\sum_{i_1,i_2,\cdots,i_k=1}^{n}\frac{\partial^k U}{\partial a_{i_1}\partial a_{i_2}\cdots\partial a_{i_k}}\left(a^0\right)\left(a_{i_1} - a^0_{\,i_1}\right)$$

$$\left(a_{i_2} - a^0_{\,i_2}\right)\cdots\left(a_{i_k} - a^0_{\,i_k}\right) \tag{18}$$

The third-order Taylor expansion formula of U is

$$U \approx U^0 + \sum_{k=1}^{3}\frac{1}{k!}\sum_{i_1,i_2,\cdots,i_k=1}^{n}\frac{\partial^k U}{\partial a_{i_1}\partial a_{i_2}\cdots\partial a_{i_k}}\left(a^0\right)\left(a_{i_1} - a^0_{\,i_1}\right)$$

$$\left(a_{i_2} - a^0_{\,i_2}\right)\cdots\left(a_{i_k} - a^0_{\,i_k}\right) \tag{19}$$

Substituting N_1 samples of interval variables and U_{n+1} (The displacement is calculated by tangent stiffness method or initial stress method (See Chapter 1 for details). U_{n+1} is the solution of Eq.1) into Eq.18 or Eq.19, the vectors $U_1, U_2, \cdots, U_{N_1}$ can be obtained.

The maximums and minimums of components of $U_1, U_2, \cdots, U_{N_1}$ are the upper bounds and lower bounds.

$$Upper_x = \max(U_{1x}, U_{2x}, \cdots, U_{N_1 x}) \tag{20}$$

where max( ) and *Upper$_x$* see chapter 4.

$$Lower_x = \min(U_{1x}, U_{2x}, \cdots, U_{N_1 x})$$ **(21)**

where min( ) and *Lower$_x$* see chapter 4.

$$Upper_y = \max(U_{1y}, U_{2y}, \cdots, U_{N_1 y})$$ **(22)**

where max( ) and *Upper$_y$* see chapter 4

$$Lower_y = \min(U_{1y}, U_{2y}, \cdots, U_{N_1 y})$$ **(23)**

where min( ) and *Lower$_y$* see chapter 4

$$Upper_z = \max(U_{1z}, U_{2z}, \cdots, U_{N_1 z})$$ **(24)**

where max( ) is and *Upper$_z$* see chapter 4.

$$Lower_z = \min(U_{1z}, U_{2z}, \cdots, U_{N_1 z})$$ **(25)**

where min( ) and *Lower$_z$* see chapter 4.

Nonlinear Interval finite element using Neumann expansion

Neumann expansion is applied to stochastic finite elements [17, 18]. Neumann expansion is also applied to interval finite elements for the nonlinear problem. Interval variable $\left[\underline{a}, \overline{a}\right]$ is generated by the following formula

$$a_i = \underline{a} + \frac{\overline{a} - \underline{a}}{n} i - \frac{\overline{a} - \underline{a}}{2} = \frac{(3n - 2i)\underline{a} + (2i - n)\overline{a}}{2n}$$ **(26)**

$$i = 1, 2, \cdots, n$$

The equilibrium equation is written as

$$KU = F \tag{27}$$

where U = the displacement vector, F = the external force, K = the global stiffness matrix.

The stiffness matrix K is decomposed into two matrices

$$K = K_0 + \Delta K \tag{28}$$

where K_0 is the stiffness matrix is replaced by midpoint values, while ΔK represents deviatoric parts. The solution U_0 can be obtained as

The Neumann expansion of K^{-1} takes the following form:

$$K^{-1} = (K_0 + \Delta K)^{-1} = (I - P + P^2 - P^3 + L)K_0^{-1} \tag{29}$$

U is represented by the following series as

$$U = U_{(0)} - U_{(1)} + U_{(2)} - U_{(3)} + \cdots \tag{30}$$

This series solution is equivalent to the following equation:

$$K_0 U_{(i)} = \Delta K U_{(i-1)} \quad i = 1, 2, \cdots \tag{31}$$

Substituting N_1 samples of interval variables and U_{n+1} into above equations, the vectors $U_1, U_2, \cdots, U_{N_1}$ can be obtained.

The maximums and minimums of component of $U_1, U_2, \cdots, U_{N_1}$ are the upper bounds and lower bounds. See section 2 for the calculation formula of upper and lower bounds of displacement.

Nonlinear Interval finite element using Sherman-Morrison- Woodbury expansion

Samples of interval variables are generated by Eq.26.

Sherman-Morrison-Woodbury expansion is expressed as

$$(K + UV^T)^{-1} = K^{-1} - K^{-1}U(I + V^T K^{-1}U)^{-1}V^T K^{-1} \tag{32}$$

We obtain

$$(K_0 + (\Delta K^T)^T)^{-1} = K_0^{-1} - K_0^{-1}\left(I + (\Delta K^T)^T K_0^{-1}\right)^{-1}(\Delta K^T)^T K_0^{-1} \tag{33}$$

Eq.2 is rewritten as

$$U = (K_0 + (\Delta K^T)^T)^{-1}F \tag{34}$$

$$= K_0^{-1}F - K_0^{-1}\left(I + (\Delta K^T)^T K_0^{-1}\right)^{-1}(\Delta K^T)^T K_0^{-1}F \tag{35}$$

Using Neumann expansion, the above formula is

$$U = (K_0 + (\Delta K^T)^T)^{-1}F$$
$$= K_0^{-1}F - K_0^{-1}\left(-(\Delta K^T)^T K_0^{-1} + ((\Delta K^T)^T K_0^{-1})^2\right) (\Delta K^T)^T K_0^{-1}F \tag{36}$$

Substituting N_1 samples of interval variables and U_{n+1} into Eq.35 or Eq.36, the vectors $U_1, U_2, \cdots, U_{N_1}$ can be obtained.

The maximums and minimums of components of $U_1, U_2, \cdots, U_{N_1}$ are the upper bounds and lower bounds. See section 2 for the calculation formula of the upper and lower bounds of displacement.

A New Iterative Method(NIM)Material properties, geometry parameters and applied loads of structures are assumed to be interval variables.

Samples of interval variables are generated by Eq.12.

Eq.2 is rewritten as

$$Ax=b, \tag{37}$$

In this section, a new iterative method (NIM) and its convergence analysis for finding a solution of (37) along with the estimation of error bounds are described. Let $A \in C^{m \times n}$ and T be a subspace of Cn. Starting with $Z0=\beta Y$, where β is a non zero realscalar,$Y \in C^{n \times m}$ satisfying R(Y) $\subseteq T$ and for any x0 $\in$ _T,the iterative method [19]is defined for k=0,1,2,... by

$$Z_{k+1}=Z_k(2I-AZ_k) \tag{38}$$

$$x_{k+1}=x_k+Z_{k+1}(b-Ax_k) \tag{39}$$

Substituting N_1 samples of interval variables and U_{n+1}into Eqs.38 and 39, the vectors $U_1,U_2,\cdots,U_{N_1}$ can be obtained. The maximums and minimums of $U_1,U_2,\cdots,U_{N_1}$ are the lower bounds and upper bounds of $U_1,U_2,\cdots,U_{N_1}$.

See section 2 for the calculation formula of the upper and lower bounds of displacement.

Elastoplastic Problem

The global equilibrium equation of the structure is the following nonlinear equations

$$\{\emptyset\} = [K(\delta)]\{\delta\} - \{F\}=0 \tag{40}$$

where $[K(\delta)]$ is global stiffness matrix, $\{\delta\}$ is displacement matrix and $\{F\}$ is a load matrix.

Material properties, geometry parameters and applied loads of structures are assumed to be interval variables. Midpoint values of the elastic modulus, Poisson's ratio ,geometry parameters and load are substituted into Eq.40.The incremental tangent stiffness method or initial stress method are used to solve displacement and stress(See Chapter 1 for details). $\{\delta\}_{n+1}$ is the solution of Eq.40 . $\{\sigma\}_{n+1}$ is the solution of stress.

After $\{\delta\}_{n+1}$ represents Eq.40, it becomes a linear equation containing interval variables of Young's modulus , Poisson's ratio ,geometry parameters and loads.

After $\{\sigma\}_{n+1}$ represents stress formula (Eq.60, Eq.80, Eq.86, Eq.91), it contains interval variables of Young's modulus , Poisson's ratio and geometry parameters.

Perturbation technology for interval finite element of the elastoplastic problem

Young's modulus , Poisson's ratio ,geometry parameters and loads are interval variables. They are expressed as

$$\bar{a} = a + \Delta a \tag{41}$$

$$\underline{a} = a - \Delta a \tag{42}$$

where a is midpoints of interval variables,Δa is a small perturbation.

Eq.40 is rewritten

$$[K]\{\delta\} = \{F\} \tag{43}$$

Using perturbation technology and representing $\{\delta\}_{n+1}$, we obtain

$$\left([K] + \Delta[K]\right)\left(\{\delta\} + \Delta\{\delta\}\right) = \left(\{F\} + \Delta\{F\}\right) \tag{44}$$

Eq.44 is rewritten

$$\left(\{\delta\} + \Delta\{\delta\}\right) = \left([K] + \Delta[K]\right)^{-1}\left(\{F\} + \Delta\{F\}\right) \tag{45}$$

Using the Neumann expansion of $\left[\tilde{K}\right]+\Delta\left[\tilde{K}\right]^{-1}$ and neglecting second-order terms, we get

$$\Delta\{\delta\}=[K]\Delta\{F\}-[K]^{-1}\Delta[K]\{\delta\} \tag{46}$$

We obtain

$$\underline{\delta} = \{\delta\}^c - \Delta\{\delta\} \tag{47}$$

$$\overline{\delta} = \{\delta\}^c + \Delta\{\delta\} \tag{48}$$

Taylor expansion for interval finite element of the elastoplastic problem

Interval variable   is generated by the following formula

$$a_i = \underline{a}+\frac{\overline{a}-\underline{a}}{n}i = \frac{i\overline{a}+(n-i)\underline{a}}{n} \tag{49}$$

$$i = 1,2,\cdots,n$$

Material properties, geometry parameters and applied loads of structures are assumed to be interval variables. They are $[\underline{a}_1,\overline{a}_1],[\underline{a}_2,\overline{a}_2],\cdots,[\underline{a}_j,\overline{a}_j],\cdots,[\underline{a}_n,\overline{a}_n]$ .The equilibrium equation is written as

$$KU = F \tag{50}$$

where U =the displacement vector, F =the external force, K =the global stiffness matrix.

By applying the Taylor series at the midpoints of the interval variables, see Section 2 for the calculation formula of $\dfrac{\partial U}{\partial a_i}$, $\dfrac{\partial^2 U}{\partial a_i \partial a_j}$, $\dfrac{\partial^3 U}{\partial a_i^2 \partial a_j}$

The Taylor expansion formula of U is

$$U = U^0 + \sum_{k=1}^{m} \frac{1}{k!} \sum_{i_1,i_2,\cdots,i_k=1}^{n} \frac{\partial^k U}{\partial a_{i_1} \partial a_{i_2} \cdots \partial a_{i_k}} \left(a^0 \right) \left(a_{i_1} - a^0_{i_1} \right)$$

$$\left(a_{i_2} - a^0_{i_2} \right) \cdots \left(a_{i_k} - a^0_{i_k} \right) +$$

$$\frac{1}{(m+1)!} \sum_{i_1,i_2,\cdots,i_{m+1}=1}^{n} \frac{\partial^{m+1} U}{\partial a_{i_1} \partial a_{i_2} \cdots \partial a_{i_k}} \left(a^0 \right) \left(a_{i_1} - a^0_{i_1} \right)$$

$$\tag{51}$$

$$\left(a_{i_2} - a^0_{i_2} \right) \cdots \left(a_{i_k} - a^0_{i_k} \right)$$

The second-order term of the Taylor expansion formula is given by

$$U \approx U^0 + \sum_{k=1}^{2} \frac{1}{k!} \sum_{i_1,i_2,\cdots,i_k=1}^{n} \frac{\partial^k U}{\partial a_{i_1} \partial a_{i_2} \cdots \partial a_{i_k}} \left(a^0 \right) \left(a_{i_1} - a^0_{i_1} \right)$$

$$\tag{52}$$

$$\left(a_{i_2} - a^0_{i_2} \right) \cdots \left(a_{i_k} - a^0_{i_k} \right)$$

The third-order Taylor expansion formula of U is

$$U \approx U^0 + \sum_{k=1}^{3} \frac{1}{k!} \sum_{i_1,i_2,\cdots,i_k=1}^{n} \frac{\partial^k U}{\partial a_{i_1} \partial a_{i_2} \cdots \partial a_{i_k}} \left(a^0 \right) \left(a_{i_1} - a^0_{i_1} \right)$$

$$\tag{53}$$

$$\left(a_{i_2} - a^0_{i_2} \right) \cdots \left(a_{i_k} - a^0_{i_k} \right)$$

Substituting N_1 samples of interval variables and U_{n+1} (The displacement is calculated by tangent stiffness method or initial stress method (See Chapter 1 for details). U_{n+1} is the solution of Eq.50) into Eq.52 or Eq.53, the vectors $U_1, U_2, \cdots, U_{N_1}$ can be obtained.

The maximums and minimums of component of $U_1, U_2, \cdots, U_{N_1}$ are the upper bounds and lower bounds.

$$Upper_x = \max(U_{1x}, U_{2x}, \cdots, U_{N_1 x}) \tag{54}$$

where max( ) and   see chapter 4.

$$Lower_x = \min(U_{1x}, U_{2x}, \cdots, U_{N_1 x}) \tag{55}$$

where min( ) and   see chapter 4.

$$Upper_y = \max(U_{1y}, U_{2y}, \cdots, U_{N_1 y}) \tag{56}$$

where max( ) and   see chapter 4.

$$Lower_y = \min(U_{1y}, U_{2y}, \cdots, U_{N_1 y}) \tag{57}$$

where min( ) and   see chapter 4.

$$Upper_z = \max(U_{1z}, U_{2z}, \cdots, U_{N_1 z}) \tag{58}$$

where max( ) is and   see chapter 4.

$$Lower_z = \min(U_{1z}, U_{2z}, \cdots, U_{N_1 z}) \tag{59}$$

where min( ) and   see chapter 4.

$$\{\sigma\} = [D]_{ep}[B]\{U\} \tag{60}$$

where $[D]_{ep}$ is elastoplastic matrix.

The stress for element d is given by

$$\{\sigma\} = [D]_{ep}[B]\{U\}^d \tag{61}$$

where $[B]$=the gradient matrix of element d and U^d =t he element d nodal displacement vector.

The incremental tangent stiffness method or initial stress method are used to solve stress(See Chapter 1 for details). $\{\sigma\}_{n+1}$ is the solution of stress. After $\{\sigma\}_{n+1}$ represents stress Eq.61, it contains interval variables of Young's modulus , Poisson's ratio and geometry parameters.

Substituting $N_1\,U^d$ and N_1 samples of interval variables into Eq.61, the vectors $\{\sigma\}_1,\{\sigma\}_2,\cdots,\{\sigma\}_{N_1}$ can be obtained.

$$Upper_{\sigma x} = \max(\sigma_{1x},\sigma_{2x},\cdots,\sigma_{N_1 x}) \tag{62}$$

where max( ) and   see chapter 4.

$$Lower_{\sigma x} = \min(\sigma_{1x},\sigma_{2x},\cdots,\sigma_{N_1 x}) \tag{63}$$

where min( ) is minimum value of x-direction component of N₁ stress vectors,and $Lower_{\sigma x}$ see chapter 4.

$$Upper_{\sigma y} = \max(\sigma_{1y},\sigma_{2y},\cdots,\sigma_{N_1 y}) \tag{64}$$

where max( ) and $Upper_{\sigma y}$ see chapter 4.

$$Lower_{\sigma y} = \min(\sigma_{1y},\sigma_{2y},\cdots,\sigma_{N_1 y}) \tag{65}$$

where min( ) and $Lower_{\sigma y}$ see chapter 4.

$$Upper_{\sigma z} = \max(\sigma_{1z},\sigma_{2z},\cdots,\sigma_{N_1 z}) \tag{66}$$

where max( ) and $Upper_{\sigma z}$ see chapter 4.

$$Lower_{\sigma z} = \min(\sigma_{1z},\sigma_{2z},\cdots,\sigma_{N_1 z}) \tag{67}$$

where min( ) and $Lower_{\sigma z}$ see. chapter 4

$$Upper_{\tau xy} = \max(\tau_{1xy}, \tau_{2xy}, \cdots, \tau_{N_1 xy}) \qquad (68)$$

where max( ) and $Upper_{\tau xy}$ see. chapter 4

$$Lower_{\tau xy} = \min(\tau_{1xy}, \tau_{2xy}, \cdots, \tau_{N_1 xy}) \qquad (69)$$

where min( ) and $Lower_{\tau xy}$ see chapter 4.

$$Upper_{\tau yz} = \max(\tau_{1yz}, \tau_{2yz}, \cdots, \tau_{N_1 yz}) \qquad (70)$$

where max( ) and $Upper_{\tau yz}$ see chapter 4.

$$Lower_{\tau yz} = \min(\tau_{1yz}, \tau_{2yz}, \cdots, \tau_{N_1 yz}) \qquad (71)$$

where min( ) and $Lower_{\tau yz}$ see chapter 4.

$$Upper_{\tau zx} = \max(\tau_{1zx}, \tau_{2zx}, \cdots, \tau_{N_1 zx}) \qquad (72)$$

where max( ) and $Upper_{\tau yz}$ see chapter 4.

$$Lower_{\tau zx} = \min(\tau_{1zx}, \tau_{2zx}, \cdots, \tau_{N_1 zx}) \qquad (73)$$

where min( ) and $Lower_{\tau zx}$ see chapter 4.

Neumann expansion for interval finite element of elastoplastic problem

Neumann expansion is applied to stochastic finite element [17, 18]. Neumann expansion is also applied to interval finite element for elastoplastic problem. Interval variable $\left[\underline{a}, \overline{a} \right]$ is generated by the following formula

$$a_i = \underline{a} + \frac{\overline{a} - \underline{a}}{n} i - \frac{\overline{a} - \underline{a}}{2} = \frac{(3n - 2i)\underline{a} + (2i - n)\overline{a}}{2n} \qquad (74)$$

$i = 1, 2, \cdots, n$

Material properties, geometry parameters and applied loads of structures are assumed to be interval variables.

The equilibrium equation is written as

$$KU = F \tag{75}$$

where U =the displacement vector, F =the external force, K =the global stiffness matrix.

The stiffness matrix K is decomposed two matrices

$$K = K_0 + \Delta K \tag{76}$$

where K_0 is the stiffness matrix is replaced by mean values, ΔK represents deviatoric parts. The solution U_0 can be obtained as

The Neumann expansion of K^{-1} takes the following form:

$$K^{-1} = (K_0 + \Delta K)^{-1} = (I - P + P^2 - P^3 + \cdots)K_0^{-1} \tag{77}$$

U is represented by the following series as

$$U = U_{(0)} - U_{(1)} + U_{(2)} - U_{(3)} + \cdots \tag{78}$$

This series solution is equivalent to the following equation:

$$K_0 U_{(i)} = \Delta K U_{(i-1)} \quad i = 1, 2, \cdots \tag{79}$$

Substituting N_1 samples of interval variables and U_{n+1} into above formulas, the vectors $U_1, U_2, \cdots, U_{N_1}$ can be obtained.

The maximums and minimums of component of $U_1, U_2, \cdots, U_{N_1}$ are the upper bounds and lower bounds. See section 2 for the calculation formula of upper and lower bounds of displacement.

The stress for element d is given by

$$\{\sigma\} = [D]_{ep}[B]\{U\}^d \tag{80}$$

The incremental tangent stiffness method or initial stress method are used to solve stress(See Chapter 1 for details). $\{\sigma\}_{n+1}$ is the solution of stress. After $\{\sigma\}_{n+1}$ represents stress Eq.80, it contains interval variables of Young's modulus , Poisson's ratio and geometry parameters.

Substituting $N_1\,U^d$ and N_1 samples of interval variables into Eq.80, the vectors $\{\sigma\}_1, \{\sigma\}_2, \cdots, \{\sigma\}_{N_1}$ can be obtained. See chapter 4 for the calculation formula of the upper and lower bounds of stress.

Nonlinear interval finite element using Sherman-Morrison- Woodbury expansion

Material properties, geometry parameters and applied loads of structures are assumed to be interval variables. Samples of interval variables are generated by Eq.74.

Sherman-Morrison-Woodbury expansion is expressed as

$$(K + UV^T)^{-1} = K^{-1} - K^{-1}U(I + V^TK^{-1}U)^{-1}V^TK^{-1} \tag{81}$$

We obtain

$$(K_0 + (\Delta K^T)^T)^{-1} = K_0^{-1} - K_0^{-1}\left(I + (\Delta K^T)^TK_0^{-1}\right)^{-1}(\Delta K^T)^TK_0^{-1} \tag{82}$$

Eq.82 is rewritten as

$$U = (K_0 + (\Delta K^T)^T)^{-1}F \tag{83}$$

$$= K_0^{-1}F - K_0^{-1}\left(I + (\Delta K^T)^T K_0^{-1}\right)^{-1}(\Delta K^T)^T K_0^{-1}F \qquad (84)$$

Using Neumann expansion, the above formula is

$$U = (K_0 + (\Delta K^T)^T)^{-1}F$$

$$= K_0^{-1}F - K_0^{-1}\left(-(\Delta K^T)^T K_0^{-1} + \left((\Delta K^T)^T K_0^{-1}\right)^2\right)(\Delta K^T)^T K_0^{-1}F \qquad (85)$$

Substituting N_1 samples of interval variables and U_{n+1} into Eq.84 or Eq.85, the vectors $U_1, U_2, \cdots, U_{N_1}$ can be obtained.

The maximums and minimums of component of $U_1, U_2, \cdots, U_{N_1}$ are the upper bounds and lower bounds. See the above section for the calculation formula of the upper and lower bounds of displacement.

The stress for element d is given by

$$\{\sigma\} = [D]_{ep}[B]\{U\}^d \qquad (86)$$

The incremental tangent stiffness method or initial stress method are used to solve stress(See Chapter 1 for details). $\{\sigma\}_{n+1}$ is thesolution of stress. After $\{\sigma\}_{n+1}$ represents stress Eq.86, it contains interval variables of Young's modulus , Poisson's ratio and geometry parameters.

Substituting N_1 U^d and N_1 samples of interval variables into Eq.86, the vectors $\{\sigma\}_1, \{\sigma\}_2, \cdots, \{\sigma\}_{N_1}$ can be obtained. The maximums and minimums of $\{\sigma\}_1, \{\sigma\}_2, \cdots, \{\sigma\}_{N_1}$ are the upper bounds and lower bounds of $\{\sigma\}_1, \{\sigma\}_2, \cdots, \{\sigma\}_{N_1}$. See chapter 4 for the calculation formula of the upper and lower bounds of stress.

The Homotopy Perturbation Method (MIHPD)

Material properties, geometry parameters and applied loads of structures are assumed to be interval variables. Samples of interval variables are generated by Eq.49.

N_1 samples of vector $\bar{a}$ are produced. N_1 matrices $\begin{bmatrix} \tilde{K} \end{bmatrix}$ and N_1 Eq.75 are generated. For nonlinear vibrations, Eq.75 is a system of nonlinear equations. A modified iteration formulas by the homotopy perturbation method (MIHPD)with accelerated fourth-and fifth-order convergence [20] is used to solve Eq.75.

Eq.75 is rewritten as

$$\Phi(X) = \begin{cases} f_1(X) \\ f_2(X) \\ \vdots \\ f_N(X) \\ X = (X_1, X_2, \cdots X_N)^T \in \mathfrak{R}^N \end{cases} \tag{87}$$

The solution of Eq.87 is given by

$$y_{i,m} = x_{i,m} - \sum_{n=1}^{N} \Psi_{i,n}(X_m) f_n(X_m) \tag{88}$$

$$z_{i,m} = -\sum_{n=1}^{N} \Psi_{i,n}(Y_m) f_n(Y_m) \tag{89}$$

$$x_{i,m+1} = y_{i,m} + z_{i,m} - \frac{1}{2} \sum_{n=1}^{N} \sum_{j=1}^{N} \sum_{k=1}^{N} \Psi_{i,n}(X_m) \frac{\partial^2 f_n(Y_m)}{\partial x_j \partial x_k} z_{j,m} z_{k,m} \tag{90}$$

$$i = 1, 2, \cdots, N, m = 0, 1, \cdots$$

Substituting N_1 samples of interval variables and U_{n+1} into Eq.75, the vectors $U_1, U_2, \cdots, U_{N_1}$ can be obtained. The maximums and minimums of $U_1, U_2, \cdots, U_{N_1}$ are the lower bounds and upper bounds of $U_1, U_2, \cdots, U_{N_1}$.

See section 2 for the calculation formula of the upper and lower bounds of displacement.

The stress for element d is given by

$$\{\sigma\} = [D]_{ep}[B]\{U\}^d$$

(91)

The incremental tangent stiffness method or initial stress method are used to solve stress(See Chapter 1 for details). $\{\sigma\}_{n+1}$ is the solution of stress. After $\{\sigma\}_{n+1}$ represents stress Eq.91, it contains interval variables of Young's modulus, Poisson's ratio and geometry parameters.

Substituting $N_1 U^d$ and N_1 samples of interval variables into Eq.91, the vectors $\{\sigma\}_1, \{\sigma\}_2, \cdots, \{\sigma\}_{N_1}$ can be obtained. The maximums and minimums of $\{\sigma\}_1, \{\sigma\}_2, \cdots, \{\sigma\}_{N_1}$ are the upper bounds and lowerbounds of $\{\sigma\}_1, \{\sigma\}_2, \cdots, \{\sigma\}_{N_1}$. See chapter 4 for the calculation formula of the upper and lower bounds of stress.

CONCLUDING REMARKS

Considering the influence of uncertain parameters on nonlinear structures, five methods of interval finite element calculation for nonlinear structures are presented. Perturbation method, Taylor expansion method, Neumann expansion, Sherman Morrison Woodbury expansion method and a new iterative method are proposed. The five calculation methods can be used for interval finite element calculation of nonlinear structures in practical engineering problems.

REFERENCES

[1]   S.S. Rao, and L. Berke, "Analysis of Uncertain Structural Systems Using Interval Analysis", *AIAA J.,* vol. 35, pp. 727-735, 1997.

http://dx.doi.org/10.2514/2.164

[2]   S. Nakagiri, and K. Suzuki, "Finite element interval analysis of external loads identified by displacement input with uncertainty", *Comput. Methods Appl. Mech. Eng.,* vol. 168, pp. 63-72, 1999.

http://dx.doi.org/10.1016/S0045-7825(98)00134-0

[3]   O. Dessombz, F. Thouverez, J-P. La^ın'e, and L. J'ez'eque, "Analysis of Mechanical Systems using Interval Computations applied to Finite Elements Methods", *J. Sound Vibrat.,* no. January, pp. 1-21, 2000.

[4]   R.L. Muhanna, and R.L. Mullen, "Uncertainty in mechanics problems-interval-based approach", *J. Eng. Mech.,* no. JUNE, pp. 557-566, 2001.

http://dx.doi.org/10.1061/(ASCE)0733-9399(2001)127:6(557)

[5]   S. Mc William, "Anti-optimisation of uncertain structures using interval analysis", *Comput. Struc.,* no. May, pp. 421-430, 2001.

http://dx.doi.org/10.1016/S0045-7949(00)00143-7

[6]   David Moens,WimDesmet,DirkVandepitte, "Interval and fuzzy dynamic analysis of finite element models with superelements", *Comput. Struc.,* vol. 85, pp. 304-319, 2007.

http://dx.doi.org/10.1016/j.compstruc.2006.10.011

[7]   A. Neumaier, and A. Pownuk, "Linear systems with large uncertainties, with applications to truss structures", *Reliab. Comput.,* vol. 13, pp. 149-172, 2007.

http://dx.doi.org/10.1007/s11155-006-9026-1

[8]   D. Degrauwe, G. Lombaert, and G. De Roeck, "Improving interval analysis in finite element calculations by means of affine arithmetic", *Comput. Struc.,* vol. 88, pp. 247-254, 2010.

http://dx.doi.org/10.1016/j.compstruc.2009.11.003

[9]   M.V. Rama Rao, R.L. Mullen, and R.L. Muhanna, *A new interval finite element formulation with the same accuracy in primary and derived variables,* .

[10] I. Elishakoff, and Y. Miglis, *"Novel parameterized intervals may lead to sharp bounds",* Mech. Res. Comm., vol. Vol. 44, Janurry, 2012, pp. 1-8.

[11] G. Muscolino, and A. Sofi, "Stochastic analysis of structures with uncertain-but-bounded parameters via improved interval analysis", *Probab. Eng. Mech.,* vol. 28, pp. 152-163, 2012.

http://dx.doi.org/10.1016/j.probengmech.2011.08.011

[12] MilanVaško,PeterPecháč, "Chosen numerical algorithms for interval finite element analysis", *Procedia Eng.,* vol. 96, pp. 400-409, 2014.

http://dx.doi.org/10.1016/j.proeng.2014.12.109

[13] R. Santoro, G. Muscolino, and I. Elishakoff, "Optimization and anti-optimization solution of combined parameterized and improved interval analyses for structures with uncertainties", *Comput. Struc.,* vol. 149, pp. 31-42, 2015.

http://dx.doi.org/10.1016/j.compstruc.2014.11.006

[14] M. Faes, J. Cerneels, D. Vandepitte, and D. Moens, "Identification and quantification of multivariate interval uncertainty in finite element models", *Comput. Methods Appl. Mech. Eng.,* vol. 315, pp. 896-920, 2017.

http://dx.doi.org/10.1016/j.cma.2016.11.023

[15] A. Sofi, E. Romeo, O. Barrera, and A. Cocks, "An interval finite element method for the analysis of structures with spatially varying uncertainties", *Adv. Eng. Softw.,* vol. 128, pp. 1-19, 2019.

http://dx.doi.org/10.1016/j.advengsoft.2018.11.001

[16] M. Faes, and D. Moens, "Multivariate dependent interval finite element analysis via convex hull pair constructions and the Extended Transformation Method", *Comput. Methods Appl. Mech. Eng.,* vol. 347, pp. 85-102, 2019.

http://dx.doi.org/10.1016/j.cma.2018.12.021

[17] F. Yamazaki, M. Shinozuka, and G. Dasgupta, "Neumann expansion for stochastic finite element analysis", *ASCE J. Engng. Mech.,* vol. 114, pp. 1335-1354, 1988.

http://dx.doi.org/10.1061/(ASCE)0733-9399(1988)114:8(1335)

[18] W. Mo, *Reliability calculations with the stochastic finite element.,* Bentham Science Publishers: Singapore, 2020.

http://dx.doi.org/10.2174/97898114855341200101

[19] S. Srivastava, and D.K. Gupta, "An iterative method for solving general restricted linear equations", *Appl. Math. Comput.,* vol. 262, pp. 344-353, 2015.

http://dx.doi.org/10.1016/j.amc.2015.04.047

[20] K. Sayevand, and H. Jafari, "On systems of nonlinear equations: some modified iteration formulas by the homotopy perturbation method with accelerated fourth-and fifth-order convergence", *Appl. Math. Model.,* vol. 40, pp. 1467-1476, 2016.

http://dx.doi.org/10.1016/j.apm.2015.06.030

[21] S. Dey, T. Mukhopadhyay, and S. Adhikari, *Uncertainty quantification in laminated composites: A meta-model based approach.,* CRC Press, 2018.

http://dx.doi.org/10.1201/9781315155593

Nonlinear Vibration Analysis of Interval Finite Element

Abstract: For the influence of non-probabilistic parameters on nonlinear vibration, the nonlinear vibration analysis of interval finite element is proposed. Using the Newmark method, nonlinear vibration is transformed into nonlinear equations. The midpoint values of interval variables are substituted into the nonlinear equations to calculate the displacement. The displacement value is substituted into the nonlinear equations, and the nonlinear equations become linear equations. Five calculation methods of interval finite element for linear vibration are extended to nonlinear vibration.

Keywords: Nonlinear vibration analysis, Newmark method, nonlinear equations, linear equations, five calculation methods, interval variable, interval finite element.

INTRODUCTION

The influence of non-probabilistic parameters on nonlinear vibration can not be ignored sometimes. The influence of non-probabilistic parameters on nonlinear vibration must be considered in some engineering problems. The membership function of fuzzy finite element is difficult to determine in engineering. Interval variables in interval finite element are relatively simple. The engineering application of interval finite element is also more convenient. An interval truncation method is proposed to obtain solutions of large amounts of uncertainty [1]. The validity of the proposed method is investigated by finite element interval analysis in a flat plate [2]. This novel interval formulation is based on a adapting Rump's algorithm for solving interval linear equations [3]. Anti-optimisation of uncertain structures is that the displacement surface produced by the uncertain parameters is monotonic [4]. The Lagrange multiplier method is applied in interval finite element [5]. The interval boundary element method is developed for considering uncertain boundary conditions [6]. The paper gives an overview of non-probabilistic finite element analysis in applied mechanics [7]. An interval-valued Sherman–Morrison–Woodbury formula is used to inverse the interval stiffness matrix [8]. Interval versus stochastic analysis of structure are derived for the bounds of the interval

Wenhui Mo

field [9]. Interval frequency response function matrix of truss structures with uncertain-but-bounded parameters is evaluated [10]. Rational Series Expansion (RSE) provides an approximate explicit expression of the frequency response functions with uncertain parameters [11]. The bounds of the interval reliability of structures under stationary stochastic excitations are evaluated [12]. Interval rational series expansion applying interval finite element analysis is proposed for the lower bound and upper bound of interval displacements and stresses [13]. A unified response surface method is thus developed for interval and stochastic finite element analysis under different uncertainty models [14]. Perturbation approach of interval finite element is used to calculate elasto-statics problems [15].The response statistics are obtained using the perturbation method of interval finite element based on a stochasticB-spline wavelet [16].

Five calculation methods of nonlinear vibration of interval finite element are presented. They are the perturbation method, Taylor expansion method, Neumann expansion, Sherman Morrison Woodbury expansion and a new iterative method. The detailed derivation of five calculation methods is given respectively.

For a nonlinear system, the dynamic equilibrium equation is given by

$$[M]\{\ddot{\delta}\}+[C]\{\dot{\delta}\}+[K]\{\delta\}=\{F\} \tag{1}$$

where $\{\ddot{\delta}\},\{\dot{\delta}\},\{\delta\}$ are the acceleration, velocity and displacement vectors. $[M],[K]$ and $[C]$ are the global mass, stiffness and damping matrices.

By using the Newmark method, Eq.1 becomes

$$\{\delta_{t+\Delta t}\}=\left[\tilde{K}(\delta_{t+\Delta t})\right]^{-1}\{\tilde{F}_{t+\Delta t}\} \tag{2}$$

where, $\{\delta_{t+\Delta t}\}$, $\left[\tilde{K}\right]$ and $\{\tilde{F}_{t+\Delta t}\}$ indicate the displacement vector, stiffness matrix and load vector at a time $t+\Delta t$.

Eq.2 is rewritten as

$$\left[\tilde{K}\left(\delta_{t+\Delta t} \right) \right]\left\{ \delta_{t+\Delta t} \right\} = \left\{ \tilde{F}_{t+\Delta t} \right\} \tag{3}$$

Material properties, geometry parameters and applied loads of structures are assumed to be interval variables. Midpoint values of the elastic modulus, Poisson's ratio ,geometry parameters and load are substituted into Eq.3.The incremental tangent stiffness method or initial stress method are used to solve displacement and stress(See Chapter 1 for details). $\{\delta\}_{n+1}$ is the solution of Eq.40 . $\{\sigma\}_{n+1}$ is the solution of stress. After $\{\delta\}_{n+1}$ represents Eq.3, it becomes a linear equation containing interval variables of Young's modulus , Poisson's ratio ,geometry parameters and loads. After $\{\sigma\}_{n+1}$ represents stress formula (Eq.10,Eq.20 ,Eq.27 ,Eq.34,Eq.39), it contains interval variables of Young's modulus , Poisson's ratio and geometry parameters . Nonlinear stochastic finite element is transformed into linear stochastic finite element.

Interval Perturbation Finite Element for Nonlinear Vibration

Material properties, geometry parameters and applied loads are assumed to be interval variables. They are expressed as

$$\bar{a} = a + \Delta a$$

$$\underline{a} = a - \Delta a \tag{4}$$

where a is midpoints of interval variables,Δa is a small perturbation.

Eq.3 is rewritten

$$\left[\tilde{K} \right]\left\{ \delta_{t+\Delta t} \right\} = \left\{ \tilde{F}_{t+\Delta t} \right\} \tag{5}$$

Using perturbation technology and representing $\{\delta\}_{n+1}$, we obtain

$$\left[\tilde{K} \right] + \Delta\left[\tilde{K} \right] \quad \delta_{t+\Delta t} \quad + \Delta \ \delta_{t+\Delta t} \quad = \quad \tilde{F}_{t+\Delta t} \quad + \Delta \ \tilde{F}_{t+\Delta t} \tag{6}$$

Using the Neumann expansion of $\left[\tilde{K}\right]+\Delta\left[\tilde{K}\right]^{-1}$ and neglecting second-order terms, we get

$$\Delta\ \delta_{t+\Delta t}=\left[\tilde{K}\right]\Delta\ \tilde{F}_{t+\Delta t}-\left[\tilde{K}\right]^{-1}\Delta\left[\tilde{K}\right]\ \delta_{t+\Delta t} \tag{7}$$

See Chapter 5 for the calculation formula of $\Delta\left\{\tilde{F}_{t+\Delta t}\right\},\Delta\ \ddot{\delta}_{t+\Delta t}\ \ ,\Delta\ \dot{\delta}_{t+\Delta t}\ $.

We obtain

$$\underline{\delta}_{t+\Delta t}=\{\delta_{t+\Delta t}\}^{c}-\Delta\{\delta_{t+\Delta t}\} \tag{8}$$

$$\overline{\delta}_{t+\Delta t}=\{\delta_{t+\Delta t}\}^{c}+\Delta\{\delta_{t+\Delta t}\} \tag{9}$$

where $\delta_{t+\Delta t}{}^{c}=\left[\tilde{K}\right]^{-1}\ \tilde{F}_{t+\Delta t}\ $, $\underline{\delta}_{t+\Delta t}\ $ is lower bound of $\ \delta_{t+\Delta t}\ $,

$\overline{\delta}_{t+\Delta t}\ $ is upper bound of $\ \delta_{t+\Delta t}$

See chapter 5 for the calculation formula of $\Delta\left\{\tilde{F}_{t+\Delta t}\right\},\Delta\ \ddot{\delta}_{t+\Delta t}\ \ ,\Delta\ \dot{\delta}_{t+\Delta t}\ $.

The calculation of $\Delta\left\{\tilde{F}_{t+\Delta t}\right\},\ \Delta\ \ddot{\delta}_{t+\Delta t}\ \ ,\Delta\ \dot{\delta}_{t+\Delta t}\ $ is for the next iteration.

Similarly, the lower bound and upper bound of vector $\left\{\delta_{t+i_1\Delta t}\right\}$ can be solved for at time $t+i_1\Delta t\left(i_1=2,3,\cdots,n_1\right)$ step by step.

At the time $t'=t+i_2\Delta t\left(i_2=1,2,\cdots,n_1\right)$, the stress for the element d is given by

$$\{\sigma\}=\left[D\right]_{ep}\left[B\right]\left\{\delta_{t'}^{d}\right\} \tag{10}$$

where $\left[B\right]$=the gradient matrix of element d , $[D]_{ep}$ is anelastoplastic matrix and $\left\{\delta_{t'}^{d}\right\}$= nodal displacement vector of element d. The incremental tangent stiffness method or initial stress method are used to solve stress(See Chapter 1 for details). $\{\sigma\}_{n+1}$ is the solution of stress. After $\{\sigma\}_{n+1}$ represents stress Eq.10, it contains interval variables of Young's modulus , Poisson's ratio and geometry parameters.

$\Delta\{\sigma\}$ is given by

$$\Delta\{\sigma\}=[D]_{ep}[B]\Delta\{\delta_{t'}^{d}\}+[D]_{ep}\Delta[B]\{\delta_{t'}^{d}\}+\Delta[D]_{ep}[B]\{\delta_{t'}^{d}\} \tag{11}$$

where $\Delta\{\sigma\}$ is the perturbation of $\{\sigma\}$

$$\{\bar{\sigma}\}=\{\sigma\}+\Delta\{\sigma\} \tag{12}$$

$$\{\underline{\sigma}\} = \{\sigma\} - \Delta\{\sigma\} \tag{13}$$

Where $\{\bar{\sigma}\}$ is upper bound of $\{\sigma\}$, $\{\underline{\sigma}\}$ is lower bound of $\{\sigma\}$.

Interval Neumann Finite Element for Nonlinear Vibration

Neumann expansion is applied to stochastic finite elements [17, 18]. Neumann expansion is also applied to interval finite elements for nonlinear vibration. Interval variable $[\underline{a},\bar{a}]$ is generated by the following formula

$$a_{i}=\underline{a}+\frac{\bar{a}-\underline{a}}{n}i-\frac{\bar{a}-\underline{a}}{2}=\frac{(3n-2i)\underline{a}+(2i-n)\bar{a}}{2n} \tag{14}$$

$$i=1,2,\cdots,n$$

Material properties, geometry parameters and applied loads of structures are assumed to be interval variables.

Neumann expansion of $\left[\tilde{K}\right]^{-1}$ takes the following form:

$$[\tilde{K}]^{-1}=([\tilde{K}_{0}]+[\tilde{K}])^{-1}=(I-P+P^{2}-P^{3}+\cdots)[\tilde{K}_{0}]^{-1} \tag{15}$$

$\{\delta_{t+\Delta t}\}$ is represented by the following series as

$$\{\delta_{t+\Delta t}\} = \{\delta_{t+\Delta t_0}\} - \{\delta_{t+\Delta t_1}\} + \{\delta_{t+\Delta t_2}\} - \{\delta_{t+\Delta t_3}\} + \cdots \tag{16}$$

This series solution is equivalent to the following equation:

$$[K_0]\{\delta_{t+\Delta t_i}\} = \Delta[K]_{i\text{-}1}\{\delta_{t+\Delta t_{i-1}}\} \quad i=1,2,\cdots \tag{17}$$

Substituting N_1 samples of interval variables and $\{\delta\}_{n+1}$ into Eqs.15,17 and 16, the vectors $N_1 \{\delta_{t+\Delta t}\}$ can be obtained.

The maximums and minimums of $N_1 \{\delta_{t+\Delta t}\}$ are the lower bounds and upper bounds of $\{\delta_{t+\Delta t}\}$. They are shown in chapter 5.

At the time $t + \Delta t$, the velocity vector and acceleration vector are obtained as

$$\{\ddot{\delta}_{t+\Delta t}\} = b_0\left(\{\delta_{t+\Delta t}\} - \{\delta_t\}\right) - b_2\{\dot{\delta}_t\} - b_3\{\ddot{\delta}_t\} \tag{18}$$

$$\{\dot{\delta}_{t+\Delta t}\} = \{\dot{\delta}_t\} + b_6\{\ddot{\delta}_t\} + b_7\{\ddot{\delta}_{t+\Delta t}\} \tag{19}$$

Then, the upper bounds and lower bounds of displacement are obtained at a time $t + i_1\Delta t \left(i_1 = 2,3,\cdots,n_1\right)$ step by step. They are shown in chapter 5. At the time $t' = t + i_2\Delta t \left(i_2 = 1,2,\cdots,n_1\right)$, the stress for element d is given by

$$\{\sigma\} = [D]_{ep}[B]\{\delta_{t'}^d\} \tag{20}$$

The incremental tangent stiffness method or initial stress method are used to solve stress (See Chapter 1 for details). $\{\sigma\}_{n+1}$ is the solution of stress. After $\{\sigma\}_{n+1}$ represents stress Eq.20, it contains interval variables of Young's modulus, Poisson's ratio and geometry parameters.

Substituting N_1 samples of random variables $a_1, a_2, \cdots, a_i, \cdots, a_{n_1}$ and N_1 $\{\delta_{t'}^d\}$ into Eq.20, the vectors $\{\sigma\}_1, \{\sigma\}_2, \cdots, \{\sigma\}_{N_1}$ can be obtained.

The maximums and minimums of $\{\sigma\}_1, \{\sigma\}_2, \cdots, \{\sigma\}_{N_1}$ are the lower bounds and upper bounds of $\{\sigma\}$. They are shown in chapter 5.

Interval Taylor Finite Element used for Nonlinear Vibration

Interval variable $\left[\underline{a}, \overline{a}\right]$ is generated by the following formula

$$a_i = \underline{a} + \frac{\overline{a} - \underline{a}}{n} i = \frac{i\overline{a} + (n-i)\underline{a}}{n} \tag{21}$$

$$i = 1, 2, \cdots, n$$

Material properties, geometry parameters and applied loads of structures are assumed to be interval variables. They are regarded as N interval variables $[\underline{a}_1, \overline{a}_1], [\underline{a}_2, \overline{a}_2], \cdots, [\underline{a}_j, \overline{a}_j], \cdots, [\underline{a}_n, \overline{a}_n]$.

At the time $t + \Delta t$, the displacement vector using the Newmark method is given by

$$\{\delta_{t+\Delta t}\} = \left[\tilde{K}\right]^{-1}\{\tilde{F}_{t+\Delta t}\} \tag{22}$$

Eq.22 can be rewritten

$$\left[\tilde{K}\right]\{\delta_{t+\Delta t}\} = \{\tilde{F}_{t+\Delta t}\} \tag{23}$$

The partial derivative of Eq.23 with respect to a_i is given by

$$\frac{\partial \{\delta_{t+\Delta t}\}}{\partial a_i} = \left[\tilde{K}\right]^{-1}\left(\frac{\partial \{\tilde{F}_{t+\Delta t}\}}{\partial a_i} - \frac{\partial \left[\tilde{K}\right]}{\partial a_i}\{\delta_{t+\Delta t}\}\right) \tag{24}$$

where, see Chapter 2 for the formula of $\dfrac{\partial \left\{ \tilde{F}_{t+\Delta t} \right\}}{\partial a_i}$.

The partial derivative of Eq.24 with respect to a_j is given by

$$\frac{\partial^2 \left\{ \delta_{t+\Delta t} \right\}}{\partial a_i \partial a_j} = \left[\tilde{K} \right]^{-1} \left(\frac{\partial^2 \left\{ \tilde{F}_{t+\Delta t} \right\}}{\partial a_i \partial a_j} - \frac{\partial \left[\tilde{K} \right]}{\partial a_i} \frac{\partial \left\{ \delta_{t+\Delta t} \right\}}{\partial a_j} \right.$$

$$\left. - \frac{\partial \left[\tilde{K} \right]}{\partial a_j} \frac{\partial \left\{ \delta_{t+\Delta t} \right\}}{\partial a_i} - \frac{\partial^2 \left[\tilde{K} \right]}{\partial a_i \partial a_j} \left\{ \delta_{t+\Delta t} \right\} \right) \tag{25}$$

where see Chapter 2 for the formula of $\dfrac{\partial^2 \left\{ \tilde{F}_{t+\Delta t} \right\}}{\partial a_i \partial a_j}$

The displacement is expanded at the midpoints of the interval variables by means of a Taylor series.

The second-order term of the Taylor expansion formula is given by

$$\left\{ \delta_{t+\Delta t} \right\} \approx \left\{ \delta_{t+\Delta t} \right\}^0 + \sum_{k=1}^{2} \frac{1}{k!} \sum_{i_1,i_2,\cdots,i_k=1}^{n} \frac{\partial^k \left\{ \delta_{t+\Delta t} \right\}}{\partial a_{i_1} \partial a_{i_2} \cdots \partial a_{i_k}} \left(a^0 \right) \left(a_{i_1} - a^0{}_{i_1} \right)$$

$$\tag{26}$$

$$\left(a_{i_2} - a^0{}_{i_2} \right) \cdots \left(a_{i_k} - a^0{}_{i_k} \right)$$

Substituting N_1 samples of interval variables and $\{\delta\}_{n+1}$ into Eq.26, the vector N_1 $\left\{ \delta_{t+\Delta t} \right\}$ can be obtained.

The maximums and minimums of components of $N_1 \left\{ \delta_{t+\Delta t} \right\}$ are the upper bounds and lower bounds of $\left\{ \delta_{t+\Delta t} \right\}$. They are shown in chapter 5.

See Chapter 5 for the calculation formula of $\dfrac{\partial\{\ddot{\delta}_{t+\Delta t}\}}{\partial a_i}$, $\dfrac{\partial\{\dot{\delta}_{t+\Delta t}\}}{\partial a_i}$ $\dfrac{\partial^2\{\ddot{\delta}_{t+\Delta t}\}}{\partial a_i\partial a_j}$ and

$\dfrac{\partial^2\{\dot{\delta}_{t+\Delta t}\}}{\partial a_i\partial a_j}$.

They must be calculated for the following iteration.

Then, the upper bounds and lower bounds of displacement are obtained at a time $t+i_1\Delta t\,(i_1=2,3,\cdots,n_1)$ step by step. They are shown in chapter 5.

At the time $t'=t+i_2\Delta t\,(i_2=1,2,\cdots,n_1)$, the stress for the element d is given by

$$\{\sigma\}=[D]_{ep}[B]\{\delta_{t'}^{d}\} \tag{27}$$

The incremental tangent stiffness method or initial stress method are used to solve stress(See Chapter 1 for details). $\{\sigma\}_{n+1}$ is the solution of stress. After $\{\sigma\}_{n+1}$ represents stress Eq.27, it contains interval variables of Young's modulus , Poisson's ratio and geometry parameters.

Substituting N_1 samples of random variables $a_1,a_2,\cdots,a_i,\cdots,a_{n_1}$ and N_1 $\{\delta_{t'}^{d}\}$ into Eq.27, the vectors $\{\sigma\}_1,\{\sigma\}_2,\cdots,\{\sigma\}_{N_1}$ can be obtained. Themaximums and minimums of $\{\sigma\}_1,\{\sigma\}_2,\cdots,\{\sigma\}_{N_1}$ are the lower bounds and upper bounds of $\{\sigma\}$. They are shown in chapter 5.

Interval Sherman-Morrison-Woodbury Expansion Finite Element

Interval variable $\left[\underline{a},\overline{a}\right]$ is generated by the following formula

$$a_i=\underline{a}+\frac{\overline{a}-\underline{a}}{n}i-\frac{\overline{a}-\underline{a}}{2}=\frac{(3n-2i)\underline{a}+(2i-n)\overline{a}}{2n} \tag{28}$$

$i=1,2,\cdots,n$

Material properties, geometry parameters and applied loads of structures are assumed to be interval variables.

At the time $t + \Delta t$, the displacement vector using the Newmark method is given by

$$\{\delta_{t+\Delta t}\} = \left[\tilde{K}\right]^{-1}\{\tilde{F}_{t+\Delta t}\} \tag{29}$$

Eq.29 is rewritten as

$$U = K^{-1}F \tag{30}$$

Using Sherman-Morrison-Woodbury expansion, we obtain

$$(K_0 + (\Delta K^T)^T)^{-1} = K_0^{-1} - K_0^{-1}\left(I + (\Delta K^T)^T K_0^{-1}\right)^{-1}(\Delta K^T)^T K_0^{-1} \tag{31}$$

$$U = (K_0 + (\Delta K^T)^T)^{-1}F = K_0^{-1}F -$$

$$K_0^{-1}\left(I + (\Delta K^T)^T K_0^{-1}\right)^{-1}(\Delta K^T)^T K_0^{-1}F \tag{32}$$

Using Neumann expansion, the above formula is

$$U = (K_0 + (\Delta K^T)^T)^{-1}F = K_0^{-1}F - K_0^{-1}\left(-(\Delta K^T)^T K_0^{-1} + \left((\Delta K^T)^T K_0^{-1}\right)^2\right)$$

$$(\Delta K^T)^T K_0^{-1}F \tag{33}$$

Substituting N_1 samples of interval variables and U_{n+1} (The displacement is calculated by tangent stiffness method or initial stress method (See Chapter 1 for details). U_{n+1} is the solution of Eq.30) into Eq.32 or Eq.33, the vectors $U_1, U_2, \cdots, U_{N_1}$ can be obtained.

The maximums and minimums of $U_1, U_2, \cdots, U_{N_1}$ are the lower bounds and upper bounds of $U_1, U_2, \cdots, U_{N_1}$. They are shown in chapter 5.

Vectors $\{\delta_{t+i_1\Delta t}\}, \{\dot{\delta}_{t+i_1\Delta t}\}, \{\ddot{\delta}_{t+i_1\Delta t}\}$ are solved at time $t+i_1\Delta t(i_1 = 2,3,\cdots,n_1)$ step-by-step. The upper bounds and lower bounds of displacement are obtained at a time $t+i_1\Delta t(i_1 = 2,3,\cdots,n_1)$ step by step. They are shown in chapter 5.

The stress for the element d is given by

$$\{\sigma\} = [D][B]\{\delta_t^d\} \tag{34}$$

The incremental tangent stiffness method or initial stress method are used to solve stress(See Chapter 1 for details). $\{\sigma\}_{n+1}$ is the solution of stress. After $\{\sigma\}_{n+1}$ represents stress Eq.34, it contains interval variables of Young's modulus , Poisson's ratio and geometry parameters.

Substituting N_1 samples of random variables $a_1, a_2, \cdots, a_i, \cdots, a_{n_1}$ and N_1 $\{\delta_t^d\}$ into Eq.34, the vectors $\{\sigma\}_1, \{\sigma\}_2, \cdots, \{\sigma\}_{N_1}$ can be obtained.

The maximums and minimums of $\{\sigma\}_1, \{\sigma\}_2, \cdots, \{\sigma\}_{N_1}$ are the lower bounds and upper bounds of $\{\sigma\}_1, \{\sigma\}_2, \cdots, \{\sigma\}_{N_1}$. They are shown inchapter 5..

The Homotopy Perturbation Method (MIHPD)

Material properties, geometry parameters and applied loads of structures are assumed to be interval variables. Samples of interval variables are generated by Eq.21.

N_1 samples of the vector $\bar{a}$ are produced. N_1 matrices $[\tilde{K}]$ and N_1 Eqs.3 are generated. For nonlinear vibrations, Eq.3 is a system of nonlinear equations. A modified iteration formula by the homotopy perturbation method (MIHPD)with accelerated fourth-and fifth-order convergence [19] is used to solve Eq.3 .

Eq.3 is rewritten as

$$\Phi(X) = \begin{cases} f_1(X) \\ f_2(X) \\ \vdots \\ f_N(X) \\ X = (X_1, X_2, \cdots X_N)^T \in \Re^N \end{cases} \tag{35}$$

The solution of Eq.35 is given by

$$y_{i,m} = x_{i,m} - \sum_{n=1}^{N} \Psi_{i,n}(X_m) f_n(X_m) \tag{36}$$

$$z_{i,m} = -\sum_{n=1}^{N} \Psi_{i,n}(Y_m) f_n(Y_m) \tag{37}$$

$$x_{i,m+1} = y_{i,m} + z_{i,m} - \frac{1}{2} \sum_{n=1}^{N} \sum_{j=1}^{N} \sum_{k=1}^{N} \Psi_{i,n}(X_m) \frac{\partial^2 f_n(Y_m)}{\partial x_j \partial x_k} z_{j,m} z_{k,m} \tag{38}$$

$$i = 1, 2, \cdots, N, m = 0, 1, \cdots$$

Substituting N_1 samples of interval variables and U_{n+1} into Eq.36, the vectors $U_1, U_2, \cdots, U_{N_1}$ can be obtained. The maximums and minimums of $U_1, U_2, \cdots, U_{N_1}$ are the lower bounds and upper bounds of $U_1, U_2, \cdots, U_{N_1}$. They are shown in chapter 5.

Vectors $\{\delta_{t+i_1\Delta t}\}, \{\dot{\delta}_{t+i_1\Delta t}\}, \{\ddot{\delta}_{t+i_1\Delta t}\}$ are solved at time $t + i_1\Delta t \, (i_1 = 2, 3, \cdots, n_1)$ step-by-step. The upper bounds and lower bounds of displacement are obtained at a time $t + i_1\Delta t \, (i_1 = 2, 3, \cdots, n_1)$ step by step. They are shown in chapter 5.

The stress for the element d is given by

$$\{\sigma\} = [D][B]\{\delta_t^d\} \tag{39}$$

The incremental tangent stiffness method or initial stress method are used to solve stress(See Chapter 1 for details). $\{\sigma\}_{n+1}$ is the solution of stress. After $\{\sigma\}_{n+1}$ represents stress Eq.39, it contains interval variables of Young's modulus , Poisson's ratio and geometry parameters.

Substituting N_1 samples of random variables $a_1, a_2, \cdots, a_i, \cdots, a_{n_1}$ and N_1 $\{\delta_{t'}^d\}$ into Eq.39, the vectors $\{\sigma\}_1, \{\sigma\}_2, \cdots, \{\sigma\}_{N_1}$ can be obtained.

The maximums and minimums of $\{\sigma\}_1, \{\sigma\}_2, \cdots, \{\sigma\}_{N_1}$ are the lower bounds and upper bounds of $\{\sigma\}_1, \{\sigma\}_2, \cdots, \{\sigma\}_{N_1}$. They are shown in chapter 5.

CONCLUDING REMARKS

The nonlinear vibration with uncertain parameters is transformed into nonlinear equations by Neumark method. Five calculation methods for linear vibration are extended to nonlinear vibration. The five calculation methods can calculate the upper and lower bounds of displacement and stress.

REFERENCES

[1]   S.S. Rao, and L. Berke, "Analysis of Uncertain Structural Systems Using Interval Analysis", *AIAA J.*, vol. 35, no. 4, pp. 727-735, 1997.
      http://dx.doi.org/10.2514/2.164

[2]   S. Nakagiri, and K. Suzuki, "Finite element interval analysis of external loads identified by displacement input with uncertainty", *Comput. Methods Appl. Mech. Eng.*, vol. 168, no. 1-4, pp. 63-72, 1999.
      http://dx.doi.org/10.1016/S0045-7825(98)00134-0

[3]   O. Dessombz, F. Thouverez, J-P. La^ın'e, and L. J'ez'eque, "Analysis of Mechanical Systems using Interval Computations applied to Finite Elements Methods", *J. Sound Vibrat.*, no. January, pp. 1-21, 2000.

[4]   R.L. Muhanna, and R.L. Mullen, "Uncertainty in mechanics problems-interval-based approach", *J. Eng. Mech.*, vol. 127, no. JUNE, pp. 557-566, 2001.
      http://dx.doi.org/10.1061/(ASCE)0733-9399(2001)127:6(557)

[5]   S. Mc William, "Anti-optimisation of uncertain structures using interval analysis", *Comput. Struc.*, vol. 79, no. May, pp. 421-430, 2001.
      http://dx.doi.org/10.1016/S0045-7949(00)00143-7

[6]   Bart F.Zalewski ,Robert L.Mullen ,Rafi L.Muhanna, "Interval boundary element method in the presence of uncertain boundary conditions, integration errors, and truncation errors",Engineering Analysis with Boundary Elements,Vol. 33, pp.508-513,April 2009, April 2009.

[7]   D. Moens, and M. Hanss, "Non-probabilistic finite element analysis for parametric uncertainty treatment in applied mechanics: Recent advances", *Finite Elem. Anal. Des.*, vol. 47, no. 1, pp. 4-16, 2011.

http://dx.doi.org/10.1016/j.finel.2010.07.010

[8]   N. Impollonia, and G. Muscolino, "Interval analysis of structures with uncertain-but-bounded axial stiffness", *Comput. Methods Appl. Mech. Eng.*, vol. 200, no. 21-22, pp. 1945-1962, 2011.
      http://dx.doi.org/10.1016/j.cma.2010.07.019

[9]   G. Muscolino, A. Sofi, and M. Zingales, "One-dimensional heterogeneous solids with uncertain elastic modulus in presence of long-range interactions: interval versus stochastic analysis", *Comput. Struc.*, vol. 122, pp. 217-229, 2013.
      http://dx.doi.org/10.1016/j.compstruc.2013.03.005

[10] G. Muscolino, and A. Sofi, "Bounds for the stationary stochastic response of truss structures with uncertain-but-bounded parameters", *Mech. Syst. Signal Process.*, vol. 37, no. 1-2, pp. 163-181, 2013.
      http://dx.doi.org/10.1016/j.ymssp.2012.06.016

[11] G. Muscolino, R. Santoro, and A. Sofi, "Explicit frequency response functions of discretized structures with uncertain parameters", *Comput. Struc.*, vol. 133, pp. 64-78, 2014.
      http://dx.doi.org/10.1016/j.compstruc.2013.11.007

[12] G. Muscolino, R. Santoro, and A. Sofi, "Reliability analysis of structures with interval uncertainties under stationary stochastic excitations", *Comput. Methods Appl. Mech. Eng.*, vol. 300, pp. 47-69, 2016.
      http://dx.doi.org/10.1016/j.cma.2015.10.023

[13] A. Sofi, and E. Romeo, "A novel Interval Finite Element Method based on the improved interval analysis", *Comput. Methods Appl. Mech. Eng.*, vol. 311, pp. 671-697, 2016.
      http://dx.doi.org/10.1016/j.cma.2016.09.009

[14] A. Sofi, and E. Romeo, "A unified response surface framework for the interval and stochastic finite element analysis of structures with uncertain parameters", *Probab. Eng. Mech.*, vol. 54, pp. 25-36, 2018.
      http://dx.doi.org/10.1016/j.probengmech.2017.06.004

[15] S. Vadlamani, and C.O. Arun, "A stochastic B-spline wavelet on the interval finite element method for problems in elasto-statics", *Probab. Eng. Mech.*, vol. 58, 2019.102996
      http://dx.doi.org/10.1016/j.probengmech.2019.102996

[16] C.O. Shashank Vadlamani, "Arun, "A stochastic B-spline wavelet on the interval finite element method for beams", *Comput. Struc.*, vol. 233, 2020.106246
      http://dx.doi.org/10.1016/j.compstruc.2020.106246

[17] F. Yamazaki, M. Shinozuka, and G. Dasgupta, "Neumann expansion for stochastic finite element analysis", *J. Eng. Mech.*, vol. 114, no. 8, pp. 1335-1354, 1988.
      http://dx.doi.org/10.1061/(ASCE)0733-9399(1988)114:8(1335)

[18] W. Mo, *Reliability calculations with the stochastic finite element.*, Bentham Science Publishers: Singapore, 2020.
      http://dx.doi.org/10.2174/97898114855341200101

[19] K. Sayevand, and H. Jafari, "On systems of nonlinear equations: some modified iteration formulas by the homotopy perturbation method with accelerated fourth-and fifth-order convergence", *Appl. Math. Model.*, vol. 40, no. 2, pp. 1467-1476, 2016.
      http://dx.doi.org/10.1016/j.apm.2015.06.030

CHAPTER 8

Random Field, Interval Field, Fuzzy Field and Mixed Field

Abstract: Material properties are assumed to be random parameters, interval parameters and fuzzy parameters. If the variation range is large, they are not assumed to be constants. Two improved methods of the random field are developed. The midpoint method, local average method, interpolation method and improved interpolation method of interval field are addressed. The midpoint method, local average method, interpolation method and improved interpolation method of the fuzzy field are presented. The calculation method of mixed field is discussed and the calculation formula is proposed.

Keywords: Stochastic field, interval field, Fuzzy field, mixed field, The midpoint method, Local average method, Interpolation method, Improved interpolation method.

INTRODUCTION

Concrete and other composite materials have spatial variability, so the material properties should not be regarded as constants, but as random processes. The material properties are assumed to be interval parameters or fuzzy parameters. If the variation range is large, it should be treated as an interval field or fuzzy field. The material properties are random, non-probabilistic and fuzzy, and should be treated as mixed field .Midpoint method for the discretization of random fields is proposed and the influence of material properties and applied loads of structures is investigated [1]. Local spatial averages efficiently evaluate the matrix of covariances [2]. Uncertain structural parameter is regarded as Gaussian stochastic process and the two-dimensional local averaging technique is extended for 3D random field [3]. The local averages method of inhomogeneous random field and non-rectangular elements is proposed using Gaussian quadrature [4]. The method that the random field is discretized is analogous to the discretization of the displacement in finite element methods [5]. A method to evaluate the stochastic fields using Karhunen- Loeve expansion is developed [6]. A weighted integral method is proposed to compute the stochastic field of material parameters [7, 8].

Wenhui Mo

An improvement of the midpoint method of stochastic field is presented [9]. The Karhunen–Loève (KL) expansion represents a random field that relies on the solution of an eigenvalue problem [10]. In order to cope with limitations on the applicability of the interval finite element analysis, the concept of interval fields is introduced [11]. The present paper determines the region of material properties and applied loads with uncertain-but-bounded parameters via interval analysis [12]. The uncertain flexibility is represented by both a random and an interval field to analyze the response variability of Euler–Bernoulli beams [13]. A method consisting of a combination of interval and random fields is proposed [14]. The improved interval analysis via extra unitary interval is proposed to solve the generalized interval eigenvalue problem [15]. A new discretization of the coupled interval field is performed to analyze the effects of Young's modulus uncertainty via interval finite element [16]. Anapproach for imprecise random field analysis using parametrized kernel functions is presented to handle imprecision [17].

Two improved methods of the random field are introduced. Four calculation methods of interval field are developed. Four calculation methods of fuzzy fields are presented. A mixed field calculation method is proposed.

Stochastic Field

The elastic modulus of the material, Poisson's ratio and the load on the structure are assumed to be random processes. If the random vector field is non-homogeneous or homogeneous, the quadrant is asymmetric, and the local average region of the random field is not rectangular. Gaussian integral is used to calculate the mean vector and covariance matrix. The elastic modulus is discussed as an example. Poisson's ratio, load, and so on.

Improved local average method

The elastic modulus of the element l is expressed as

$$b_l(x) = \frac{\iiint_{v_l} b(x)}{v_l} \tag{1}$$

where v_l is the volume of element l, $b(x)$ is a random process representing the elastic modulus and $b_l(x)$ is the elastic modulus of element l

After coordinate transformation, the above formula is written in Jacobi form. The above formula is converted to Gaussian numerical integration.

The elastic modulus of element l is expressed as

$$b_l(x) = \frac{\sum_{k=1}^{n}\sum_{j=1}^{n}\sum_{i=1}^{n} W_i W_j W_k b_l\left(\zeta_i, \eta_j, \xi_k\right)}{V_l} \tag{2}$$

where w is the weight coefficient of the one-dimensional Gaussian integral. ζ_i, η_j, ξ_k are the integration points. n is the number of integral points in each coordinate direction.

The mean value of elastic modulus of the element   is

$$E(b_l(x)) = \frac{\sum_{k=1}^{n}\sum_{j=1}^{n}\sum_{i=1}^{n} W_i W_j W_k \overline{b}_l\left(\zeta_i, \eta_j, \xi_k\right)}{V_l} \tag{3}$$

where $\overline{b}_l\left(\zeta_i, \eta_j, \xi_k\right)$ is the mean of the elastic modulus of the Gaussian integral points , $E(\ )$ is the mean..

The covariance of elastic modulus of element   and element m is

$$\text{cov}\left(b_l(x), b_m(x)\right) =$$

$$\frac{\sum_{k'=1}^{n}\sum_{j'=1}^{n}\sum_{i'=1}^{n} \sum_{k=1}^{n}\sum_{j=1}^{n}\sum_{i=1}^{n} W_{i'} W_{j'} W_{k'} W_i W_j W_k \, \text{cov}(b_l\left(\zeta_i, \eta_j, \xi_k\right), b_m\left(\zeta_i, \eta_j, \xi_k\right))}{V_l V_m} \tag{4}$$

where cov( ) is the covariance, V_m is the volume of element m.

Improved Interpolation Method

The stochastic process c (x) is approximately expressed as a shape function

$$c(x) = \sum_{i=1}^{q} N_i(x)c_i \tag{5}$$

where $N_i(x)$ is the shape function, c_i is the elastic modulus of the element node, and c(x) is the random process representing the elastic modulus.

Using the local average method, the elastic modulus of the element   is expressed as

$$c_l(x) = \frac{\iiint_{v_l} c(x)}{v_l} \tag{6}$$

where v_l is the volume of element l, $c(x)$ is a random process representing the elastic modulus and . c$_l$(x) is the elastic modulus of element l.

After coordinate transformation, the above formula is written in Jaccobi form. The above formula is converted to Gaussian numerical integration.

The elastic modulus of the element l is expressed as

$$c_l(x) = \frac{\sum_{k=1}^{n}\sum_{j=1}^{n}\sum_{i=1}^{n} W_i W_j W_k c_l\left(\zeta_i, \eta_j, \xi_k\right)}{V_l} \tag{7}$$

where W is the weight coefficient of the one-dimensional Gaussian integral. ζ_i, η_j, ξ_k are the integration points. n is the number of integral points in each coordinate direction. $c_l\left(\zeta_i, \eta_j, \xi_k\right)$ is the elastic modulus of the Gaussian integral points.

The mean value of elastic modulus of the element   is

$$E(c_l(x)) = \frac{\sum_{k=1}^{n}\sum_{j=1}^{n}\sum_{i=1}^{n} W_i W_j W_k \overline{c_l}\left(\zeta_i, \eta_j, \xi_k\right)}{V_l} \tag{8}$$

where $\overline{c_l}\left(\zeta_i,\eta_j,\xi_k\right)$ is the mean of the elastic modulus of the Gaussian integral points, $E(\ )$ is the mean.

The covariance of elastic modulus of element l and element m is

$$\operatorname{cov}\left(c_l\left(x\right),c_m\left(x\right)\right)=$$

$$\frac{\sum_{k'=1}^{n}\sum_{j'=1}^{n}\sum_{i'=1}^{n}\sum_{k=1}^{n}\sum_{j=1}^{n}\sum_{i=1}^{n}W_iW_{j'}W_{k'}W_iW_jW_k\operatorname{cov}(c_l\left(\zeta_i,\eta_j,\xi_k\right),c_m\left(\zeta_i,\eta_j,\xi_k\right))}{V_lV_m} \qquad (9)$$

where cov( ) is the covariance, V_m is the volume of element m.

Interval Field

The elastic modulus of the material, Poisson's ratio and the load on the structure are assumed to be interval variables. The elastic modulus is discussed as an example. Poisson's ratio, load, and so on.

Midpoint method

The upper bound of the elastic modulus of the element l is $\overline{b_l}=\overline{b}_{mid}$ $\qquad$ (10)

where $\overline{b_l}$ is the upper bound of the elastic modulus of the element l, $\overline{b}_{mid}$ is the upper bound of the elastic modulus at the midpoint of the element l.

The lower bound of the elastic modulus of the element l is

$$\underline{b_i}=\underline{b}_{mid} \qquad\qquad (11)$$

where $\underline{b_i}$ is the lower bound of the elastic modulus of element l, $\underline{b}_{mid}$ is the lower bound of the elastic modulus at the midpoint of the element l.

Local integration method

The upper bound of the elastic modulus of the element l is

$$\overline{b}_l = \frac{\sum_{k=1}^{n}\sum_{j=1}^{n}\sum_{i=1}^{n} W_i W_j W_k \overline{b}_l\left(\varsigma_i,\eta_j,\xi_k\right)}{V_l} \tag{12}$$

Where $\overline{b}_l$ is the upper bound of the elastic modulus of the element l.

The lower bound of the elastic modulus of the element l is

$$\underline{b}_l = \frac{\sum_{k=1}^{n}\sum_{j=1}^{n}\sum_{i=1}^{n} W_i W_j W_k \underline{b}_l\left(\varsigma_i,\eta_j,\xi_k\right)}{V_l} \tag{13}$$

where $\underline{b}_i$ is the lower bound of the elastic modulus of the element ,

Interpolation method

The upper bound of the elastic modulus of the element   is

$$\overline{c}(x) = \sum_{i=1}^{q} N_i(x)\overline{c}_i \tag{14}$$

where $\overline{c}(x)$ is the upper bound of the elastic modulus of the element l. $\overline{c}_i$ is the upper bound of the elastic modulus of the element node for element l.

The lower bound of the elastic modulus of the element l is

$$\underline{c}(x) = \sum_{i=1}^{q} N_i(x)\underline{c}_i \tag{15}$$

where $\underline{c}_i$ is the lower bound of the elastic modulus of the element node for element l.

Improved interpolation method

The upper bound of the elastic modulus of the element *l* is

$$\bar{c}_l(x) = \frac{\sum_{k=1}^{n}\sum_{j=1}^{n}\sum_{i=1}^{n} W_i W_j W_k \bar{c}_l\left(\zeta_i, \eta_j, \xi_k\right)}{V_l} \tag{16}$$

where $\bar{c}(x)$ is the upper bound of the elastic modulus of the element *l*.

The lower bound of the elastic modulus of the element *l* is

$$\underline{c}_l(x) = \frac{\sum_{k=1}^{n}\sum_{j=1}^{n}\sum_{i=1}^{n} W_i W_j W_k \underline{c}_l\left(\zeta_i, \eta_j, \xi_k\right)}{V_l} \tag{17}$$

where $\underline{c}_i(x)$ is the lower bound of the elastic modulus of the element l.

Fuzzy Field

The elastic modulus of the material, Poisson's ratio and the load on the structure are assumed to be fuzzy variables and fuzzy processes. The elastic modulus is discussed as an example. Poisson's ratio, load, and so on. The membership function of elastic modulus is $\mu_B(x)$.

α-cut is defined as

$$B_\alpha = \left\{ x \in X, \mu_B(x) \geq \alpha \right\} \tag{18}$$

For the ithα-cut, the lower and upper bounds are given by

$$\underline{b} = \min\{b: b \in B_\alpha\} \tag{19}$$

$$\bar{b} = \max\{b: b \in B_\alpha\} \tag{20}$$

If a set of n α-cut levels set is constructed, n lower and upper bounds are given by

$$B = \left\{ (\underline{b}, \overline{b})_{\alpha_1}, (\underline{b}, \overline{b})_{\alpha_2}, \cdots, (\underline{b}, \overline{b})_{\alpha_n} \right\} \tag{21}$$

Midpoint method

The upper bound of the elastic modulus of the element *l* is

$$\overline{b}_{l_\alpha} = \overline{b}_{mid_\alpha} \tag{22}$$

Where $\overline{b}_{l_\alpha}$ is the upper bound of the elastic modulus of element *l* for α-cut, $\overline{b}_{mid_\alpha}$ is the upper bound of the elastic modulus at the midpoint of element *l* for α-cut.

The lower bound of the elastic modulus of element   for α-cut is

$$\underline{b}_{i_\alpha} = \underline{b}_{mid_\alpha} \tag{23}$$

where $\underline{b}_{i_\alpha}$ is the lower bound of the elastic modulus of element *l* for α-cut, $\underline{b}_{mid_\alpha}$ is the lower bound of the elastic modulus at the midpoint of element *l* for α-cut.

Local integration method

The upper bound of the elastic modulus of element   for α-cut is

$$\overline{b}_{l_\alpha} = \frac{\sum_{k=1}^{n}\sum_{j=1}^{n}\sum_{i=1}^{n} W_i W_j W_k \overline{b}_{l_\alpha}\left(\zeta_i, \eta_j, \xi_k\right)}{V_l} \tag{24}$$

Where $\overline{b}_{l_\alpha}$ is the upper bound of the elastic modulus of element *l* for α-cut.

The lower bound of the elastic modulus of element   for α-cut is

$$\underline{b}_{l_\alpha} = \frac{\sum_{k=1}^{n}\sum_{j=1}^{n}\sum_{i=1}^{n} W_i W_j W_k \underline{b}_{l_\alpha}\left(\zeta_i, \eta_j, \xi_k\right)}{V_l} \tag{25}$$

where $\underline{b}_{l_\alpha}$ is the lower bound of the elastic modulus of element l forα-cut.

Interpolation method

The upper bound of the elastic modulus of element   forα-cut is

$$\overline{c}_\alpha(x) = \sum_{i=1}^{q} N_i(x)\overline{c}_{i_\alpha} \tag{26}$$

where $\overline{c}_\alpha(x)$ is the upper bound of the elastic modulus of element *l* forα-cut. $\overline{c}_{i_\alpha}$ is the upper bound of the elastic modulus of the element node of the element *l*. for α-cut.

The lower bound of the elastic modulus of element   forα-cut is

$$\underline{c}_\alpha(x) = \sum_{i=1}^{q} N_i(x)\underline{c}_{i_\alpha} \tag{27}$$

where $\underline{c}_\alpha(x)$ is the lower bound of the elastic modulus of element *l* forα-cut, $\underline{c}_{i_\alpha}$ is the lower bound of the elastic modulus of the element node of element *l* forα-cut.

Improved interpolation method

The upper bound of the elastic modulus of element   forα-cut is

$$\overline{c}_{l_\alpha}(x) = \frac{\sum_{k=1}^{n}\sum_{j=1}^{n}\sum_{i=1}^{n} W_i W_j W_k \overline{c}_{l_\alpha}\left(\varsigma_i, \eta_j, \xi_k\right)}{V_l} \tag{28}$$

where $\overline{c}_{l_\alpha}(x)$ is the upper bound of the elastic modulus of element *l* forα-cut.

The lower bound of the elastic modulus of element *l* forα-cut is

$$\underline{c}_{l_\alpha}(x) = \frac{\sum\limits_{k=1}^{n}\sum\limits_{j=1}^{n}\sum\limits_{i=1}^{n} W_i W_j W_k \underline{c}_{l_\alpha}\left(\zeta_i, \eta_j, \xi_k\right)}{V_l} \tag{29}$$

Where $\underline{c}_{l_\alpha}(x)$ is the lower bound of the elastic modulus of element l for α-cut,

Mixed Field

The elastic modulus of the material, Poisson's ratio and the load on the structure are assumed to be stochastic variables, interval variables, fuzzy variables. They obey random processes, interval processes and fuzzy processes. The elastic modulus is discussed as an example. Poisson's ratio, load, and so on. The membership function of elastic modulus is $\mu_F(x)$.

When random variables obey arbitrary distribution, the samples are generated by the following method.

$$P\left\{|x - \mu'| \geq \varepsilon\right\} \leq \frac{\sigma^2}{\varepsilon^2} \tag{30}$$

where μ' is the mean, σ is the standard deviation, ε is any positive number, and the above formula is called Chebyshev inequality.

To take $\varepsilon = 10\sigma_i$, $x = a_i$, we get

$$P\left\{|a_i - \mu_i| < 10\sigma_i\right\} \geq 0.99 \tag{31}$$

then

$$a_i < 10\sigma_i + \mu_i \tag{32}$$

and

$$a_i > -10\sigma_i + \mu_i \tag{33}$$

The random field adopts the midpoint method, local average method, interpolation method, improved local average method and improved interpolation method to solve mean and standard deviation respectively , the upper and lower bounds of the elastic modulus of the element   are expressed as $\overline{b}_{i_s}$, $\underline{b}_{i_s}$.

$$\overline{b}_{i_s} = 9.99\sigma_i + \mu_i \tag{34}$$

$$\underline{b}_{i_s} = -9.99\sigma_i + \mu_i \tag{35}$$

The upper bound of the elastic modulus of element *l* is

$$\overline{b}_l = \overline{b}_{l_s} + \overline{b}_{l_q} + \overline{b}_{l_\alpha} \tag{36}$$

Where $\overline{b}_l$ is the upper bound of the elastic modulus of the element *l* , $\overline{b}_{l_s}$ is the upper bound of the elastic modulus of element *l* using the midpoint method, local average method, interpolation method, improved local average method and improved interpolation method of stochastic field, respectively. $\overline{b}_{l_q}$ is the upper bound of the elastic modulus of the element *l* using the midpoint method, interpolation method, improved local average method and improved interpolation method of interval field respectively. $\overline{b}_{l_\alpha}$ is the upper bound of the elastic modulus of the element *l* using the midpoint method, interpolation method, improved local average method and improved interpolation method of fuzzy field respectively forα-cut.

The lower bound of the elastic modulus of the element *l* is

$$\underline{b}_l = \underline{b}_{l_s} + \underline{b}_{l_q} + \underline{b}_{l_\alpha} \tag{37}$$

where $\underline{b}_l$ is the lower bound of the elastic modulus of the element *l*. $\underline{b}_{l_s}$ is the lower bound of the elastic modulus of the element *l* using the midpoint method, local average method, interpolation method, improved local average method and improved interpolation method of stochastic field respectively. $\underline{b}_{l_q}$ is the lower bound of the elastic modulus of the element *l* using the midpoint method, interpolation method, improved local average method and improved interpolation method of interval field respectively. $\underline{b}_{l_\alpha}$ is the lower bound of the elastic modulus of the element l using the midpoint method, interpolation method, improved local average method and improved interpolation method of fuzzy field respectively for α-cut.

CONCLUDING REMARKS

Considering the spatial variability of materials, two new calculation methods of the random field are proposed. When the range of interval variables is large, four interval field calculation methods are proposed. When the fuzzy parameters of materials change greatly, four calculation methods of the fuzzy field are presented. When the material parameters are random, non-probabilistic and fuzzy, a mixed field calculation method is formulated.

REFERENCES

[1] A. Der Kiureghian, and J-B. Ke, "The stochastic finite element method in structural reliability", *Probab. Eng. Mech.,* vol. 3, pp. 83-91, 1988.

http://dx.doi.org/10.1016/0266-8920(88)90019-7

[2] E. Vanmarcke, M,Grigoriu, *"Stochastic finite element analysis of simple beams",J Engrg Mech.,* vol. Vol. 109, Desember, 1983, pp. 1203-1214.

[3] S. Chakraborty, and B. Bhattacharyya, *"An efficient 3D stochastic finite element method",* Solids and *Struct.,* vol. Vol. 39, Desember, 2002, pp. 2465-2475.

[4] W.Q. Zhu, Y.J. Ren, and W.Q. Wu, "Stochastic FEM based on local averages of random vector fields", *J Engng Mech,ASC E,* vol. 118, pp. 496-511, 1992.

http://dx.doi.org/10.1061/(ASCE)0733-9399(1992)118:3(496)

[5] W K Liu, T Belytschko, and A. Mani, *Random field finite element.*

http://dx.doi.org/10.1002/nme.1620231004

[6] G. Deodatis, "Bounds on response variability of stochastic finite element systems", *J Engng Mech ASCE,* vol. 116, pp. 565-585, 1990.

http://dx.doi.org/10.1061/(ASCE)0733-9399(1990)116:3(565)

[7] M. Shinozuka, and G. Deodatis, "Response Variability of stochastic finite element systems", *J Engng Mech ASCE,* vol. 114, pp. 499-519, 1988.

http://dx.doi.org/10.1061/(ASCE)0733-9399(1988)114:3(499)

[8] C.C. Li, *Der kiureghian A., "Optimal discretizatiom of random fields",* J Engng Mech ASCE., vol. Vol. 119, Desember, 1993, pp. 1136-1154.

[9] W. Mo, "Stochastic finite element for structural vibration", *Math. Probl. Eng.,* no. September, pp. 22-40, 2010.

[10] W. Betz, I. Papaioannou, and D. Straub, "Numerical methods for the discretization of random fields by means of the Karhunen-Loève expansion", *Comput. Methods Appl. Mech. Eng.,* vol. 271, pp. 109-129, 2014.

http://dx.doi.org/10.1016/j.cma.2013.12.010

[11] W. Verhaeghe, W. Desmet, D. Vandepitte, and D. Moens, "Interval fields to represent uncertainty on the output side of a static fe analysis", *Comput. Methods Appl. Mech. Eng.,* vol. 260, pp. 50-62, 2013.

http://dx.doi.org/10.1016/j.cma.2013.03.021

[12] G. Muscolino, and A. Sofi, "Bounds for the stationary stochastic response of truss structures with uncertain-but-bounded parameters", *Mech. Syst. Signal Process.,* vol. 37, pp. 163-181, 2013.

http://dx.doi.org/10.1016/j.ymssp.2012.06.016

[13] A. Sofi, "Structural response variability under spatially dependent uncertainty: stochastic versus interval model", *Probab. Eng. Mech.,* vol. 42, pp. 78-86, 2015.

http://dx.doi.org/10.1016/j.probengmech.2015.09.001

[14] M. Faes, G.D. Sabyasachi, and D. Moens, *"Hybrid spatial uncertainty analysis for the estimation of imprecise failure probabilities in laser sintered pa-12 parts",* Comput. Math. Appl., vol. Vol. 78, Desember, 2019, pp. 2395-2406.

[15] A. Sofi, G. Muscolino, and I. Elishakoff, "Natural frequencies of structures with interval parameters", *J. Sound Vibrat.,* vol. 347, pp. 79-95, 2015.

http://dx.doi.org/10.1016/j.jsv.2015.02.037

[16] A. Sofi, G. Muscolino, and I. Elishakoff, "Static response bounds of timoshenko beams with spatially varying interval uncertainties", *Acta Mech.,* vol. 226, pp. 3737-3748, 2015.

http://dx.doi.org/10.1007/s00707-015-1400-9

[17] M. Faes, and D. Moens, "Imprecise random field analysis with parametrized kernel functions", *Mech. Syst. Signal Process.,* vol. 134, no. December, 2019.106334

http://dx.doi.org/10.1016/j.ymssp.2019.106334

CHAPTER 9

Mixed Finite Element

Abstract: The parameters of the structure contain random variables and interval variables. The Taylor expansion method and Neumann expansion method of random interval finite element are proposed. The parameters of the structure are random and fuzzy. Taylor expansion method and Neumann expansion method of the random fuzzy finite element are illustrated. The parameters of the structure are random, fuzzy and non-probabilistic. The mixed finite element calculation should be carried out using Taylor expansion and Neumann expansion.

Keywords: Mixed finite element, Neumann expansion, Random interval finite element, Random fuzzy finite element, Random fuzzy and interval finite element, Taylor expansion.

INTRODUCTION

Finite element method deals with deterministic engineering problems. The influence of uncertain factors must be considered. Uncertain factors influence the strength and life of a structure. Many engineering problems should consider uncertain properties of material, geometry and loads. The structure is affected by two or three uncertain factors, and the structure should be calculated by the mixed finite element method.

Several types of stochastic finite element methods exist in the literature :the Monte Carlo Simulation (MCS) [1-5] , the perturbation method [2, 6-9] and the spectral stochastic finite element method [10-13]. According to first-order or second-order perturbationmethods, calculation formulas of perturbation stochastic finite element method (PSFEM) are derived [2, 6-8]. Finite element solutions for material variability can be obtained by means of perturbation stochastic finite element [2]. A major advantage of perturbation stochastic finite element is that the multivariate distribution function need not be known [6]. The vibration equation of a system is transformed into a static problem by using the Newmark method and the Taylor expansion [7]. Considering the influence of random factors, sensitivity computation for a linear vibration is illustrated [8, 9]. The PSFEM is an adequate tool for nonlinear structural dynamics [14]. This paper presents a framework for probability

Wenhui Mo

sensitivity estimation of a class of problems involving linear stochastic finite element models [15]. Fuzzy finite element analysis based on the theory of fuzzy sets is presented to take account of the uncertainty of the elastic modulus and Poisson's ratio [16]. An elastoplastic finite element analysis with fuzzy parameters is proposed using fuzzy mathematics [17]. A fuzzy finite-element approach for vibration analysis involving vagueness is developed [18]. Finite element analysis of flexible multibody systems with fuzzy parameters is presented to predict the dynamic response and evaluate the sensitivity [19]. Neumann expansion for fuzzy finite element analysis can resolve the uncertain eigenvalue problem with fuzzy parameters [20].A unified respons esurface framework for interval and stochastic finite element analysis of structures with both probabilistic and non-probabilistic parameters is developed [21]. Modifications for the fuzzy and fuzzy–stochastic FEM is proposed [22]. The static analysis of structures subjected to uncertain loads using the fuzzy and intervalfinite element method is investigated [23].

The calculation formula of random interval finite element is investigated using Taylor expansion and Neumann expansion. The stochastic fuzzy finite element method is developed. Two calculation methods of mixed finite elements are presented.

Stochastic and Interval Finite Element

The elastic modulus , Poisson's ratio of the material and the load on the structure are assumed to be random process and interval variables. The global equilibrium equation of the structure is the following linear equations

$$[K]U = \{F\} \tag{1}$$

The elastic modulus , Poisson's ratio and loads of the structure are regarded as n random variables $a_1, a_2, \cdots, a_i, \cdots, a_n$ and n interval variables $[\underline{b}_1, \overline{b}_1], [\underline{b}_2, \overline{b}_2], \ldots, [\underline{b}_j, \overline{b}_j], \cdots, [\underline{b}_n, \overline{b}_n]$

Taylor expansion method

The partial derivative of Eq.1 with respect to a_i is given by

$$\frac{\partial U}{\partial a_i} = [K]^{-1} \left(\frac{\partial \{F\}}{\partial a_i} - \frac{\partial [K]}{\partial a_i} U \right) \tag{2}$$

where $\dfrac{\partial U}{\partial a_i}$ is the partial derivative of U with respect to a_i.

The partial derivative of Eq.2 with respect to a_j is given by

$$\frac{\partial^2 U}{\partial a_i \partial a_j} = [K]^{-1} \left(\frac{\partial^2 \{F\}}{\partial a_i \partial a_j} - \frac{\partial [K]}{\partial a_i} \frac{\partial U}{\partial a_j} - \frac{\partial [K]}{\partial a_j} \frac{\partial U}{\partial a_i} - \frac{\partial^2 [K]}{\partial a_i \partial a_j} U \right) \tag{3}$$

where $\dfrac{\partial^2 U}{\partial a_i \partial a_j}$ is the partial derivative of $\dfrac{\partial U}{\partial a_i}$ with respect to a_j. The displacement is expanded at the mean points of the random variables, and the mean value is taken on both sides, we get

$$E\{U\} \approx \{U\}\big|_{a=\bar{a}} + \frac{1}{2} \sum_{i=1}^{n} \sum_{j=1}^{n} \frac{\partial^2 \{U\}}{\partial a_i \partial a_j}\Big|_{a=\bar{a}} Cov(a_i, a_j) \tag{4}$$

The variance of $\{U\}$ can be calculated by the following formula:

$$Var\{U\} \approx \sum_{i=1}^{n} \sum_{j=1}^{n} \frac{\partial \{U\}}{\partial a_i}\Big|_{a=\bar{a}} \cdot \frac{\partial \{U\}}{\partial a_j}\Big|_{a=\bar{a}} \cdot Cov(a_i, a_j) \tag{5}$$

where $Var\{U\}$ is the variance of U.

Chebyshev inequality can be rewritten as

$$P\{|x - \mu'| < \varepsilon\} \geq 1 - \frac{\sigma^2}{\varepsilon^2} \tag{6}$$

where, μ' is the mean, σ is the standard deviation, ε is any positive number.

To take $\varepsilon = 100\sigma_i$, $x = U$, we get

$$P\left\{\left|U - \mu_i\right| < 100\sigma_i\right\} \geq 0.9999 \tag{7}$$

then

$$U < 100\sigma_i + \mu_i \tag{8}$$

and

$$U > -100\sigma_i + \mu_i \tag{9}$$

The upper and lower bounds of U are defined as .

$$\bar{U} \approx 99.999\sigma_i + \mu_i \tag{10}$$

$$\underline{U} \approx -99.999\sigma_i + \mu_i \tag{11}$$

where $\bar{U}$ *and* $\underline{U}$ are the upper and lower bounds of U.

.Interval variable $\left[\underline{b}, \bar{b}\right]$ is generated by the following formula

$$b_i = \underline{b} + \frac{\bar{b} - \underline{b}}{n}i = \frac{i\bar{b} + (n-i)\underline{b}}{n} \tag{12}$$

$$i = 1, 2, \cdots, n$$

Material properties, geometry parameters and applied loads of structure are assumed to be interval variables. They are $[\underline{b}_1, \bar{b}_1], [\underline{b}_2, \bar{b}_2], \cdots, [\underline{b}_j, \bar{b}_j], \cdots, [\underline{b}_n, \bar{b}_n]$.

By applying Taylor series at midpoints of interval variables, the following equations are given by

$$U^0 = \left([K]^0\right)^{-1}\{F\}^0 \tag{13}$$

$$\frac{\partial U}{\partial b_i} = [K]^{-1} \left(\frac{\partial \{F\}}{\partial b_i} - \frac{\partial [K]}{\partial b_i} U \right)$$

(14)

$$\frac{\partial^2 U}{\partial b_i \partial b_j} = [K]^{-1} \left(\frac{\partial^2 \{F\}}{\partial b_i \partial b_j} - \frac{\partial [K]}{\partial b_i} \frac{\partial U}{\partial b_j} - \frac{\partial [K]}{\partial b_j} \frac{\partial U}{\partial b_i} - \frac{\partial^2 [K]}{\partial b_i \partial b_j} U \right)$$

(15)

The second-order term of the Taylor expansion formula is given by

$$U \approx U^0 + \sum_{k=1}^{2} \frac{1}{k!} \sum_{i_1,i_2,\cdots,i_k=1}^{n} \frac{\partial^k U}{\partial b_{i_1} \partial b_{i_2} \cdots \partial b_{i_k}} \left(b^0 \right) \left(b_{i_1} - b^0_{i_1} \right)$$

(16)

$$\left(b_{i_2} - b^0_{i_2} \right) \cdots \left(b_{i_k} - b^0_{i_k} \right)$$

Substituting N_1 samples of interval variables into Eq.16, the vectors $U_1, U_2, \cdots, U_{N_1}$ can be obtained.

The N_1 displacement values were obtained by interval finite element using Taylor expansion. After the computer program is used to compare, the upper and lower bounds of U can be obtained.

$$Upper_x = \max(U_{1x}, U_{2x}, \cdots, U_{N_1 x})$$

(17)

where max( ) and $Upper_x$ see chapter 4.

$$Lower_x = \min(U_{1x}, U_{2x}, \cdots, U_{N_1 x})$$

(18)

where min( ) and $Lower_x$ see chapter 4.

$$Upper_y = \max(U_{1y}, U_{2y}, \cdots, U_{N_1 y})$$

(19)

where max( ) and $Upper_y$ see chapter 4.

$$Lower_y = \min(U_{1y}, U_{2y}, \cdots, U_{N_1 y})$$

(20)

where min( ) and *Lower*$_y$ see chapter 4..

$$Upper_z = \max(U_{1z}, U_{2z}, \cdots, U_{N_1z}) \tag{21}$$

where max( ) is and *Upper*$_z$ see chapter 4..

$$Lower_z = \min(U_{1z}, U_{2z}, \cdots, U_{N_1z}) \tag{22}$$

where min( ) and *Lower*$_z$ see chapter 4.

Upper bound of U_{mt} can be obtained

$$\overline{U}_{mt} = \overline{U}_{St} + \overline{U}_{It} \tag{23}$$

where U_{mt} is mixed displacement, $\overline{U}_{mt}$ is the upper bound of mixed displacement, $\overline{U}_{St}$ is the upper bound of displacement using stochastic finite element(Eq.10), $\overline{U}_{It}$ is upper bound of displacement using interval finite element.

Lower bound of U_{mt} can be obtained

$$\underline{U}_{mt} = \underline{U}_{St} + \underline{U}_{It} \tag{24}$$

where U_{mt} is mixed displacement, $\underline{U}_{mt}$ is lower bound of mixed displacement, $\underline{U}_{St}$ is lower bound of displacement using stochastic finite element(Eq.11), $\underline{U}_{It}$ is lower bound of displacement using interval finite element.

Neumann Expansion Method

The elastic modulus, Poisson's ratio of the material and the load on the structure are assumed to be random process and interval variables. The midpoint method, local average method and interpolation method of random field and the improved local average method and improved interpolation method of the random field in the previous chapter is adopted respectively to obtain covariance matrix of elastic modulus.

Using covariance matrix, the correlation of elastic modulus between any two elements is given by

$$C_{aa} = \begin{pmatrix} Cov(a_1,a_1) & Cov(a_1,a_2) & \cdots & Cov(a_1,a_N) \\ Cov(a_2,a_1) & Cov(a_2,a_2) & \cdots & Cov(a_2,a_N) \\ \vdots & \vdots & & \vdots \\ Cov(a_N,a_1) & Cov(a_N,a_2) & \cdots & Cov(a_N,a_N) \end{pmatrix} \tag{25}$$

A vector $\bar{a} = [a_1, a_2, \cdots, a_N]^T$ is generated [24]

$$\bar{a} = LZ \tag{26}$$

$Z = [Z_1, Z_2, \cdots, Z_N]^T$ consists of N Gaussian random variables with mean zero and unit standard deviation. The Cholesky matrix L can be obtained through a decomposition of the covariance matrix, therefore

$$\mu \lfloor ZZ^T \rfloor = I \tag{27}$$

$$LL^T = C_{aa} \tag{28}$$

I is the identity matrix. The generation of the vector $\bar{a}$ must satisfy the covariance matrix

$$\mu \lfloor \bar{a}\bar{a}^T \rfloor = \mu \lfloor LZ(LZ)^T \rfloor$$

$$= L\mu \lfloor ZZ^T \rfloor L^T = C_{aa} \tag{29}$$

Once the decomposition has been completed, different samples of the vector $\bar{a}$ can be acquired easily by Eq.26 [24].

The Neumann expansion of $[K]^{-1}$ takes the following form:

$$[K]^{-1}=([K]^0+\Delta[K])^{-1}=(I-P+P^2-P^3+\cdots)([K]^0)^{-1} \tag{30}$$

U is represented by the following series as[24]

$$U=U_{(0)}-U_{(1)}+U_{(2)}-U_{(3)}+\cdots \tag{31}$$

This series solution is equivalent to the following equation [24]:

$$[K]^0U_{(i)}=\Delta[K]U_{(i-1)}\ i=1,2,\cdots \tag{32}$$

Substituting N_1 samples of interval variables into above equations, the vectors $U_1,U_2,\cdots,U_{N_1}$ can be obtained.

The elastic modulus, Poisson's ratio of the material and the load on the structure are assumed to be interval variables. The Neumann expansion method of interval finite element can be seen in chapter 4. We obtain

$$U_m=U_S+U_I \tag{33}$$

where U_m is mixed displacement, U_S is the N displacement values obtained by the Neumann expansion method of stochastic finite element. U_I is the N displacement values obtained by the Neumann expansion method of interval finite element.

After the computer program is used to compare , the upper and lower bounds of U can be obtained.

The above calculation is based on the example that the elastic modulus is stochastic. If the elastic modulus, Poisson's ratio, geometry parameters and load are stochastic variables and interval variables, the same method can be used for calculation.

Stochastic and Fuzzy Finite Element

The elastic modulus, Poisson's ratio of the material and the load on the structure are assumed to be random processes and fuzzy processes. The global equilibrium equation of the structure is the following linear equations

$$KU = F \tag{34}$$

The elastic modulus , Poisson's ratio and loads of the structure are regarded as n random variables $a_1, a_2, \cdots, a_i, \cdots, a_n$ and n fuzzy variables..

Taylor Expansion Method

Taylor expansion method of the stochastic finite element can be seen in the previous section.

The membership function of elastic modulus is $\mu_B(x)$. The membership function of Poisson's ratio is $\mu_{F_1}(x)$. The membership function of loads is $\mu_{F_2}(x)$.

α-cut is defined as

$$B_\alpha = \left\{ x \in X, \mu_B(x) \geq \alpha \right\} \tag{35}$$

For the ithα-cut, the lower and upper bounds are given by

$$\underline{b} = min\{b : b \in B_\alpha\}$$

$$\overline{b} = max\{b : b \in B_\alpha\}$$

If a set of n α-cut levels set is constructed, n lower and upper bounds are given by

$$B = \left\{ \left(\underline{b}, \overline{b} \right)_{\alpha_1}, \left(\underline{b}, \overline{b} \right)_{\alpha_2}, \cdots, \left(\underline{b}, \overline{b} \right)_{\alpha_n} \right\} \tag{36}$$

$\left[\underline{b}, \overline{b} \right]$ is generated for aα-cut level by the following formula

$$b_i = \underline{b} + \frac{\overline{b} - \underline{b}}{n} i = \frac{i\overline{b} + (n-i)\underline{b}}{n} \tag{37}$$

$$i = 1, 2, \cdots, n$$

By applying the Taylor series at midpoints of interval variables, the following equations are given by

$$U^0 = \left(K^0\right)^{-1} F^0 \tag{38}$$

$$\frac{\partial U}{\partial b_i} = \left(K^0\right)^{-1} \left(\frac{\partial F}{\partial b_i} - \frac{\partial K}{\partial b_i} U^0 \right) \tag{39}$$

$$\frac{\partial^2 U}{\partial b_i \partial b_j} = \left(K^0\right)^{-1} \left(\frac{\partial^2 F}{\partial b_i \partial b_j} - \frac{\partial^2 K}{\partial b_i \partial b_j} U^0 - \frac{\partial K}{\partial b_i} \frac{\partial U}{\partial b_j} - \frac{\partial K}{\partial b_j} \frac{\partial U}{\partial b_i} \right) \tag{40}$$

$$\frac{\partial^3 U}{\partial b_i^2 \partial b_j} = (K^0)^{-1} \left(\frac{\partial^3 F}{\partial b_i^2 \partial b_j} - \frac{\partial^3 K}{\partial b_i^2 \partial b_j} U^0 - 3 \frac{\partial^2 K}{\partial b_i \partial b_j} \frac{\partial U}{\partial b_i} - 3 \frac{\partial K}{b_i} \frac{\partial^2 U}{\partial b_i \partial b_j} \right) \tag{41}$$

The second-order term of the Taylor expansion formula is given by

$$U \approx U^0 + \sum_{k=1}^{2} \frac{1}{k!} \sum_{i_1,i_2,\cdots,i_k=1}^{n} \frac{\partial^k U}{\partial b_{i_1} \partial b_{i_2} \cdots \partial b_{i_k}} \left(b^0\right)\left(b_{i_1} - b^0{}_{i_1}\right)$$

$$\left(b_{i_2} - b^0{}_{i_2}\right) \cdots \left(b_{i_k} - b^0{}_{i_k}\right) \tag{42}$$

The third-order Taylor expansion formula of U is

$$U \approx U^0 + \sum_{k=1}^{3} \frac{1}{k!} \sum_{i_1,i_2,\cdots,i_k=1}^{n} \frac{\partial^k U}{\partial b_{i_1} \partial b_{i_2} \cdots \partial b_{i_k}} \left(b^0\right)\left(b_{i_1} - b^0{}_{i_1}\right)$$

$$\left(b_{i_2} - b^0{}_{i_2}\right) \cdots \left(b_{i_k} - b^0{}_{i_k}\right) \tag{43}$$

Substituting N_1 samples of interval variables into Eq.42 or Eq.43, the vectors $U_1, U_2, \cdots, U_{N_1}$ can be obtained. The N_1 displacement values are obtained by interval finite element using Taylor expansion. After the computer program is used to compare, the upper and lower bounds of U can be obtained .

For a α-cut level, we obtain

Upper bound of U_{mf} can be obtained

$$\bar{U}_{mf} = \bar{U}_{Sf} + \bar{U}_{B_\alpha f}$$ **(44)**

where U_{mf} is mixed displacement, $\bar{U}_{mf}$ is the upper bound of mixed displacement, $\bar{U}_{Sf}$ is the upper bound of displacement using Taylor stochastic finite element(Eq.10), $\bar{U}_{B_\alpha f}$ is upper bound of displacement using Taylor fuzzy finite element.

Lower bound of U_{mf} can be obtained

$$\underline{U}_{mf} = \underline{U}_{Sf} + \underline{U}_{B_\alpha f}$$ **(45)**

where $\underline{U}_{mf}$ is lower bound of mixed displacement, $\underline{U}_{Sf}$ is lowerbound of displacement using Taylor stochastic finite element(Eq.11), $\underline{U}_{B_\alpha f}$ is lower bound of displacement using Taylor fuzzy finite element.

For a set of N_2 α-cut levels, we obtain $N_2 \cdot \bar{U}_{mf}$, $N_2 \underline{U}_{mf}$.

After the computer program is used to compare, the upper and lower bounds of U can be obtained.

The above calculation is based on the example that the elastic modulus is fuzzy. If the elastic modulus, Poisson's ratio and load are stochastic and fuzzy, the same method can be used for calculation.

Neumann Expansion Method

The Neumann expansion method of the stochastic finite element can be seen in the previous section.

$[\underline{b}, \bar{b}]$ is generated for aα-cut level by the following formula

$$b_i = \underline{b} + \frac{\overline{b} - \underline{b}}{n} i - \frac{\overline{b} - \underline{b}}{2} = \frac{3\underline{b} - \overline{b}}{2} + \frac{\overline{b} - \underline{b}}{n} i \tag{46}$$

$$i = 1, 2, \cdots, n$$

The Neumann expansion of K^{-1} takes the following form:

$$K^{-1} = (K^0 + \Delta K)^{-1} = (I - P + P^2 - P^3 + \cdots)(K^0)^{-1} \tag{47}$$

U is represented by the following series as

$$U = U_{(0)} - U_{(1)} + U_{(2)} - U_{(3)} + \cdots \tag{48}$$

This series solution is equivalent to the following equation:

$$K^0 U_{(i)} = \Delta K U_{(i-1)} \quad i = 1, 2, \cdots \tag{49}$$

Substituting N_1 samples of interval variables into above equations, the vectors $U_1, U_2, \cdots, U_{N_1}$ can be obtained.

.We obtain

$$U_{mn} = U_{Sn} + U_{B_\alpha n} \tag{50}$$

Where U_{mn} is mixed displacement, U_{Sn} is the N displacement values obtained by the Neumann expansion method of stochastic finite element. $U_{B_\alpha n}$ is the displacement values obtained by the Neumann expansion method of fuzzy finite element for a α-cut level.

For a set of N3 α-cut levels, we obtain N3 $U_{B_\alpha n}$.

After the computer program is used to compare , the upper and lower bounds of U can be obtained.

Stochastic, Interval and Fuzzy Finite Element

The elastic modulus , Poisson's ratio of the material and the load on the structure are assumed to be a random process ,interval process and fuzzy processes. The elastic modulus , Poisson's ratio and loads of the structure are regarded as n random variables, n interval variables and n fuzzy variables..

For a α-cut level, we obtain

Upper bound of U_{m1} can be obtained

$$\overline{U_{m1}} = \overline{U}_{S1} + \overline{U}_{I1} + \overline{U}_{B_\alpha 1} \tag{51}$$

where U_{m1} is mixed displacement, $\overline{U}_{m1}$ is upper bound of mixed displacement, $\overline{U}_{S1}$ is upper bound of displacement using Taylor stochastic finite element(Eq.10), $\overline{U}_{I1}$ is the upper bound of displacement using Taylor interval finite element, $\overline{U}_{B_\alpha 1}$ is upper bound of displacement using Taylor fuzzy finite element.

Lower bound of U_{m1} can be obtained

$$\underline{U}_{m1} = \underline{U}_{S1} + \underline{U}_{I1} + \underline{U}_{B_\alpha 1} \tag{52}$$

where U_{m1} is mixed displacement, $\underline{U}_{m1}$ is lower bound of mixed displacement, $\underline{U}_{S1}$ is lower bound of displacement using Taylor stochastic finite element(Eq.11), $\underline{U}_{I1}$ is lower bound of displacement using Taylor interval finite element $\underline{U}_{B_\alpha 1}$ is lower bound of displacement using Taylor fuzzy finite element.

For a set of N_2 α-cut levels, we obtain N_2. $\overline{U}_{m1}$, N_2 $\underline{U}_{m1}$.

After the computer program is used to compare , the upper and lower bounds of U can be obtained.

For a α-cut level, we obtain

Upper bound of U_{m2} can be obtained

$$\overline{U_{m2}} = \overline{U}_{S2} + \overline{U}_{I2} + \overline{U}_{B_\alpha 2} \tag{53}$$

where$U_{m2}$is mixed displacement, $\overline{U}_{m2}$ is upper bound of mixed displacement, $\overline{U}_{S2}$ is upper bound of displacement using Neumannstochastic finite element, $\overline{U}_{I2}$ is the upper bound of displacement using Neumann interval finite element,$\overline{U}_{B_\alpha 2}$ is upper bound of displacement using Neumann fuzzy finite element.

Lower bound of U_{m2} can be obtained

$$\underline{U}_{m2} = \underline{U}_{S2} + \underline{U}_{I2} + \underline{U}_{B_\alpha 2} \tag{54}$$

where $\underline{U}_{m2}$ is lower bound of mixed displacement, $\underline{U}_{S2}$ is lower bound of displacement using Neumann stochastic finite element, $\underline{U}_{I2}$ is lower bound of displacement using Neumann interval finite element. $\underline{U}_{B_\alpha 2}$ is lower bound of displacement using Neumann fuzzy finite element.

For a set of N_3 α-cut levels, we obtain N_3 $\overline{U}_{m2}$, N_3 $\underline{U}_{m2}$.

After the computer program is used to compare , the upper and lower bounds of U can be obtained.

CONCLUDING REMARKS

The structure is affected by random, non-probabilistic and fuzzy factors. It is necessary to calculate the structure by the mixed finite element method. Using Taylor expansion and Neumann expansion, random interval finite element, random fuzzy finite element and random interval fuzzy finite element are developed.

REFERENCES

[1]   J. Astill, C.J. Nosseir, and M. Shinozuka, "Impact loading on structures with random properties", *J Struct. Mech,* vol. 1, no. 1, pp. 63-67, 1972.
       http://dx.doi.org/10.1080/03601217208905333
[2]   F. Yamazaki, M. Shinozuka, and G. Dasgupta, "Neumann expansion for stochastic finite element analysis", *J. Eng. Mech.,* vol. 114, no. 8, pp. 1335-1354, 1988.
       http://dx.doi.org/10.1061/(ASCE)0733-9399(1988)114:8(1335)

[3] M. Papadrakakis, and V. Papadopoulos, "Robust and efficient methods for stochastic finite element analysis using Monte Carlo Simulation", *Comput. Methods Appl. Mech. Eng.*, vol. 134, no. 3-4, pp. 325-340, 1996.

http://dx.doi.org/10.1016/0045-7825(95)00978-7

[4] D.C. Charmpis, and M. Papadrakakis, "Improving the computational efficiency in finite element analysis of shells with uncertain properties", *Comput. Methods Appl. Mech. Eng.*, vol. 19, no. 4, pp. 1447-1478, 2005.

http://dx.doi.org/10.1016/j.cma.2003.12.075

[5] D.C. Charmpis, "Incomplete factorization preconditioners for the iterative solution of stochastic finite element equations", *Comput. Struc.*, vol. 88, no. 3-4, pp. 178-188, 2010.

http://dx.doi.org/10.1016/j.compstruc.2009.09.010

[6] W.K. Liu, T. Belytschko, and A. Mani, "Mani A. Random field finite element", *Int. J. Numer. Methods Eng.*, vol. 23, no. 10, pp. 1831-1845, 1986.

http://dx.doi.org/10.1002/nme.1620231004

[7] W. Mo, "Stochastic finite element for structural vibration", *Math. Probl. Eng.*, no. September, pp. 22-40, 2010.

[8] W. Mo, "Dynamic analysis of stochastic finite element based on sensitivity computation",

[9] W. Mo, "Stochastic finite element of structural vibration based on sensitivity analysis", *2010 International conference on Mechanical Engineering and Green Manufacturing,* 2010pp. 25-32.

http://dx.doi.org/10.4028/www.scientific.net/AMM.34-35.25

[10] S. Acharjee, and N. Zabaras, "Uncertainty propagation in finite deformations—a spectral stochastic Lagrangian approach", *Comput. Methods Appl. Mech. Eng.*, vol. 195, no. 19-22, pp. 2289-2312, 2006.

http://dx.doi.org/10.1016/j.cma.2005.05.005

[11] MF Ngah, and A Young, *Application of the spectral stochastic finite element method for performance prediction of composite structures, .*

http://dx.doi.org/10.1016/j.compstruct.2005.11.009

[12] S.P. Oliveira, and J.S. Azevedo, "Spectral element approximation of Fredholm integral eigenvalue problems", *J. Comput. Appl. Math.*, vol. 257, pp. 46-56, 2014.

http://dx.doi.org/10.1016/j.cam.2013.08.016

[13] P. Zakian, and N. Khaji, "A novel stochastic-spectral finite element method for analysis of elastodynamic problems in the time domain", *Meccanica,* vol. 51, no. 4, pp. 893-920, 2016.

http://dx.doi.org/10.1007/s11012-015-0242-9

[14] W.K. Liu, T. Belytschko, and A. Mani, "Probabilistic finite elements for nonlinear structural dynamics", *Comput. Methods Appl. Mech. Eng.*, vol. 57, no. 1, pp. 61-81, 1986.

http://dx.doi.org/10.1016/0045-7825(86)90136-2

[15] A. Marcos, *Valdebenito,Herman B.Hernandez,Hector A.Jensen, "Probabilitysensitivity estimation of linear stochastic finite element models applying line sampling",Strnctural safety.*, vol. Vol. 81, Desember, 2019.

[16] S. Valliappan, and T.D. Pham, "Fuzzy finite element analysis of a foundation on elastic soil medium", *Int. J. Numer. Anal. Methods Geomech.*, vol. 17, no. 11, pp. 771-789, 1993.

http://dx.doi.org/10.1002/nag.1610171103

[17] S. Valliappan, and T.D. Pham, "Elasto-plastic finite element analysis with fuzzy parameters", *Int. J. Numer. Methods Eng.*, vol. 38, no. 4, pp. 531-548, 1995.

http://dx.doi.org/10.1002/nme.1620380403

[18] L. Chen, and S.S. Rao, "Fuzzy finite-element approach for the vibration analysis of imprecisely-defined systems", *Finite Elem. Anal. Des.,* vol. 27, no. 1, pp. 69-83, 1997.
http://dx.doi.org/10.1016/S0168-874X(97)00005-X

[19] M. Tamer, "Wasfy, Ahmed K. Noor, "Finite element analysis of flexible multibody systems with fuzzy parameters", *Comput. Methods Appl. Mech. Eng.,* vol. 160, no. 3-4, pp. 223-243, 1998.
http://dx.doi.org/10.1016/S0045-7825(97)00297-1

[20] B. Lallemand, G. Plessis, T. Tison, and P. Level, "Neumann expansion for fuzzy finite element analysis", *Eng. Comput.,* vol. 16, no. 5, pp. 572-583, 1999.
http://dx.doi.org/10.1108/02644409910277933

[21] A. Sofi, and E. Romeo, "A unified response surface framework for the interval and stochastic finite element analysis of structures with uncertain parameters", *Probab. Eng. Mech.,* vol. 54, pp. 25-36, 2018.
http://dx.doi.org/10.1016/j.probengmech.2017.06.004

[22] D. Pivovarov, K. Willner, and P. Steinmann, "On spectral fuzzy–stochastic FEM for problems involving polymorphic geometrical uncertainties", *Comput. Methods Appl. Mech. Eng.,* vol. 350, pp. 432-461, 2019.
http://dx.doi.org/10.1016/j.cma.2019.02.024

[23] S. Diptiranjan Behera, "Chakraverty, "Solving the nondeterministic static governing equations of structures subjected to various forces under fuzzy and interval uncertainty", *Int. J. Approx. Reason.,* vol. 116, pp. 43-61, 2020.
http://dx.doi.org/10.1016/j.ijar.2019.10.011

[24] F. Yamazaki, M. Shinozuka, and G. Dasgupta, "Neumann expansion for stochastic finite element analysis", *J. Eng. Mech.,* vol. 114, no. 8, pp. 1335-1354, 1988.
http://dx.doi.org/10.1061/(ASCE)0733-9399(1988)114:8(1335)

SUBJECT INDEX

A

Advanced Monte Carlo methods 24
Affine arithmetic 64, 98
Analysis 2, 3, 24, 53, 63, 65, 67, 69, 70, 71,
    73, 75, 79, 80, 98, 99, 120, 148
  buckling 2
  cohesive interface 3
  dynamic 53, 98
  elasto-statics 80
  intuitionistic fuzzy fault tree 53
  nonlinear finite element 3
  non-probabilistic finite element 79, 120
  perturbation-based finite element method 24
  probabilistic 24
  probabilistic stability 24
  sensitivity 98
  static 63, 65, 67, 69, 70, 71, 73, 75, 148
  uncertainty propagation 99
Anti-optimization solution 64
Applications 1, 2, 34, 79, 98, 120
  engineering 79, 120
Approach 3, 53, 64, 79, 99, 148
  finite-element 148
  flexible 64
  fuzzy maximum entropy 53
  nonlinear finite element 3
  non-probabilistic 79
  response surface 99

B

Bernoulli beams 135
Biodegradation 2
Bridge construction 24

C

Calculation 7, 18, 53, 76, 123, 154, 157
  static 76

Calculation methods 99, 121, 134, 135, 145,
    148
  derivation of five 99, 121
Calculation process 64
Capacity assessment 3
Carbon-nanotube-reinforced-polymer (CNRP)
    24
Cholesky decomposition 32
Cholesky matrix 153
Composite facet shell element 2
Computational 2, 64
  algorithms 64
  elasto-plasticity 2
Computerized model 53
Computer program 55, 56, 151, 154, 156, 157,
    158, 159, 160
Concrete structures 2, 52
  reinforced 2
Convergence 2, 15, 35, 43, 47, 52, 76, 95,
    106, 116, 130
  accelerated fourth-and fifth-order 43, 47,
    116, 130
  analysis 35, 76, 95, 106
  properties 2
Covariance 134, 135, 136, 138, 152, 153
  matrix 134, 135, 152, 153
  of elastic modulus of element 136, 138

D

Damping mechanisms 2
Dams, concrete gravity 2
Derivation processes 79
DFSMC systems 53
Displacement vectors 36, 65, 66, 67, 70, 89,
    92, 94, 101, 104, 108, 111, 113, 121,
    126
Distribution probability density functions 32

E

Earthquakes 1, 24

non-stationary 24
Elastic 1, 3, 12, 15
  analysis 15
  composites 1
  plastic problem 1, 3, 12
Elastic modulus 4, 12, 15, 43, 135, 136, 137,
      138, 139, 140, 141, 142, 143, 144, 148,
      152, 154, 155, 157, 159
  atthemidpoint 141
  of element 135, 136, 137, 138, 141, 142,
      İ43, 144
  membership function of 140, 143, 155
  value of 4, 12, 15, 43, 136, 137
Elastoplastic 14, 18, 98, 106, 107, 108, 112,
      148
  problems 14, 18, 98, 106, 107, 108, 112
Element-by-element technique 79
Element method 24, 52, 147, 160
  conditional random finite 24
  mixed finite 147, 160
  spectral 52
Equations 4, 12, 15, 54, 86, 96, 104, 154, 158
  linear differential 96
Equilibrium equation 3, 15, 25, 35, 59, 64, 70,
      80, 99, 101, 104, 106, 108, 113, 121,
      148, 154
  dynamic 35, 59, 80, 121
  global 3, 99, 106, 148, 154

F

Finite element 1, 2, 3, 17, 18, 24, 25, 52, 53,
      54, 63, 80, 120, 147, 148
  interval analysis 120
  method 1, 2, 3, 17, 18, 24, 25, 52, 53, 54,
      63, 80, 147, 148
Finite element analysis 2, 18, 24, 52, 53, 64,
      79, 80, 98, 99, 121, 148
  method 18
  of complex structures 52
  of flexible multibody systems 148
  of structures 80, 148
Finite element calculation 53, 63, 64, 117, 147
  mixed 147

three-dimensional 53
Finite element reliability analyses of 24
  nonlinear frame structures 24
Finite element reliability 23, 24, 25
  analysis 24
  calculation 25
  methods 23
Finite elements models 2, 3, 52
  damped 52
  nonlinear 2, 3
Flexible multibody systems 148
Framework 3, 147
  computational 3
Frequency response function (FRF) 24, 52,
      63, 79, 121
  analysis 63
Friction-based devices 24
Function 5, 32, 45, 58, 79, 136, 137, 147
  multivariate distribution 147
  probability density 32, 45, 58, 79
Fuzzy 52, 53, 58, 63, 98, 120, 134, 135, 140,
      143, 145, 147, 148, 154, 157, 158, 159
  data 52
  event 58
  field 134, 135, 140, 145
  Finite Element 154, 159
  Lambda 52
  mathematics 148
  methods 53
  processes 140, 143, 154, 159
Fuzzy logic 52, 53
  control algorithm 52
  method 52
Fuzzy reliability 52, 53, 59, 60
  algorithm 53
  analysis 53
  calculation 52, 54, 59
  method 53
  of nonlinear Structures 60

G

Galerkin method, element-free 23
Gaussian 134, 135, 136, 137, 138, 153

quadrature 134
stochastic process 134
Geometric properties 2

H

Homotopy perturbation method 23, 25, 43, 47, 116, 130

I

Influence of interval variables 80, 98
Initial stress 1, 5, 99, 107, 109, 111, 114, 115, 117, 122, 124, 125, 128, 129, 130, 132
  method 1, 99, 107, 109, 111, 114, 115, 117, 122, 124, 125, 128, 129, 130, 132
  value 5
Integration points 136, 137
Interpolation method 134, 139, 142, 144, 145, 152
Interval 52, 63, 79, 80, 120, 135, 145, 159
  analysis 63, 79, 80, 135
  and fuzzy finite element 52, 63, 159
  boundary element methods 79, 120
  equations 79
  field calculation methods 135, 145
Interval 89, 120, 122, 126, 143, 159
  perturbation finite element for nonlinear vibration 122
  processes 143, 159
  stiffness matrix 120
  Taylor finite element 89, 126
  truncation method 120
Interval finite element 63, 64, 76, 99
  analysis 64
  methods of 63, 76, 99
Interval Neumann finite element 85
  for linear vibration 85
  for nonlinear vibration 124
Intuitionistic fuzzy 53
  failure rates 53
  fault-tree, based 53
Iterative 4, 79
  algorithm 79

formula 4

J

Jaccobi form 137

K

Karhunen- Loeve expansion 134

L

Lagrange multiplier method 79, 98, 120
Laminated composite 2
  plates 2
  spherical shell panel 2
Linear vibration problem 96
Loads 1, 36, 46, 54, 82, 121, 148
  random 1
  static 54
  uncertain 148
  vector 36, 46, 82, 121
Local 2, 134, 138, 137, 141, 144, 152
  average method 2, 134, 137, 144, 152
  integration method 138, 141
  spatial averages 134
Lower bounds 75, 76, 86, 87, 91, 92, 93, 94, 95, 105, 106, 114, 115, 117, 125, 128, 130, 131, 132, 150
  of displacement 75, 76, 92, 94, 95, 105, 106, 114, 115, 117

M

Markov 3
  chain Monte Carlo technique 3
  diffusion theory 3
Material 3, 4, 5, 12, 99, 135, 138, 140, 143, 145, 147, 148, 152, 154, 159
  softening 3
  stress-strain 3, 99
Mathematical 9, 11, 58
  expectations 58

treatment 9, 11
Matrix 4, 13, 14, 16, 25, 44, 65, 79, 85, 101,
    104, 108, 110, 113, 123, 153
  anelastoplastic 123
  elastic 13, 14
  elastic-plastic 13
  elastoplastic 16, 44, 110
  plastic 14
  tangent elasticity 4
Methods, inequality-based 79
Modified iteration 23, 25, 43, 47, 116, 130
    formulas 23, 43, 47, 116, 130
    method 25
Monte Carlo method 99
Monte Carlo simulation (MCS) 1, 2, 23, 25,
    34, 35, 42, 43, 44, 47, 48, 147
    method 35
Morrison-Woodbury expansion 63

N

Neumann 11, 160
  fuzzy 160
  interval 160
  Stochastic Finite Element 11
Neumann expansion 63, 70, 71, 83, 85, 98, 99,
    100, 103, 104, 105, 112, 113, 124, 147,
    148
  for interval finite element of elastoplastic
    problem 112
  of interval finite element for static analysis
    63
Neumann expansion method 147, 154, 158,
  of fuzzy finite element 158
  of interval finite element 154
  of random interval 147
  of stochastic finite element 154, 158
Neural networks 53
Newmark method 79, 80, 85, 89, 92, 94, 96,
    120, 121, 126, 129
Nodal displacement vector 28, 40, 68, 85, 87,
    123
Nonlinear 1, 2, 3, 9, 11, 23, 24, 25, 48, 120,
    132

behaviour 2
  equations 3, 9, 11, 120, 132
  finite elements 1, 24
  proxy 3
  ships rolling 3
  static problems 23, 25, 48
Nonlinear stochastic 1, 2, 3, 5, 7, 9, 11, 13, 15,
    17, 18, 122
  problems 2
Non-probabilistic parameters 120, 148
  influence of 120
Numerical 63, 98, 136, 137
  algorithms 98
  analysis method 63
  integration 136, 137

P

Parameters 23, 33, 55, 56, 80, 147
  geometric 23, 33, 55, 56
Parametrized kernel functions 135
Perturbation 1, 2, 23, 79, 80, 85, 98, 99, 117,
    121, 124, 147
  method 1, 23, 79, 80, 98, 99, 117, 121, 147
  second-order 1
  technique 98
Perturbation stochastic 24, 147
  finite element method (PSFEM) 24, 147
Perturbation technology 1, 9, 25, 82, 99, 100,
    107, 122
  for nonlinear interval finite element 99
Petri nets and fuzzy Lambda 52
Plane strain compression (PSC) 52
Poisson's ratio 4, 7, 8, 12, 15, 43, 107, 122,
    135, 138, 140, 143, 148, 154, 155, 155
  and loads 148, 155, 159
Poisson's ratiogeometry parameters 15
Polynomials 2
Preconditioned stochastic Krylov subspace 2
Probabilistic risk assessment 24
Problems 3, 52, 53
  analyzed heat conduction 53
  nonlinear magnetostatic 52
  nonlinear mechanical 3

Proxy finite element analysis (PFEA) 3

Q

Quadratic response surface method 23
Qualitative data processing 53

R

Random 2, 4, 11, 12, 15, 17, 24, 33, 92, 94,
    95, 134, 135, 137, 143, 147, 148, 149,
    152, 154, 159, 160
  interval fuzzy 160
  material properties 2
  processes 2, 3, 4, 8, 9, 11, 12, 15, 17, 24,
    33, 92, 94, 95, 134, 135, 137, 143, 147,
    148, 149, 152, 154, 159
  properties 24
  system properties 2
  variables 3, 4, 8, 9, 11, 12, 15, 17, 33, 92,
    94, 95, 147, 148, 149
Rational series expansion (RSE) 121
Reinforced concrete haunched beams 3
Relaxation iteration method 56
Relaxed directional method 53
Reliability 3, 23, 24, 32, 34, 35, 42, 43, 45, 47,
    48, 52
  approach 24
  deriving 24
  design 32
  dynamic 23, 24
  methods 23
  of linear structures 24
  of Nonlinear Structures 43
  of stiffness 32, 35, 42, 43, 47
  sensitivity analysis 3, 24
Reliability assessment 24
  lifetime 24
Reliability calculation 23, 24, 25, 27, 29, 31,
    33, 34, 35, 37, 39, 41, 43, 45, 46, 47, 48
  methods 23, 48
  of nonlinear vibration 46
  of static problems 25
Response 79, 148

dynamic 148
stationary stochastic 79
Rump's algorithm, adapting 120

S

Second order reliability method 25, 44, 48
  for nonlinear static problems 25
Sensitivity computation 147
Sherman-Morrison-Woodbury expansion 63,
    64, 74, 93, 105, 114, 129
  method 64
Stationary stochastic excitations 121
Stiffness 25, 32, 35, 36, 42, 43, 47, 59, 65, 80,
    101, 104, 108, 113, 121
  global 25, 65, 101, 104, 108, 113
Stiffness matrix 3, 4, 6, 11, 13, 36, 46, 70, 71,
    81, 99, 104, 106, 113, 121
  elastic 13
  global 4, 70, 99, 106
  initial tangent 6
  inversion 81
  tangent 4
Stochastic 1, 2, 3, 7, 17, 18, 23, 24, 25, 35, 52,
    54, 79, 134, 135, 136, 143, 148, 152,
    154, 157
  adaptive 2
  and fuzzy finite element 154
  and interval finite element 148
  field 134, 135
  finite element and structural reliability
    calculation methods 23
  fuzzy 148
  perturbation-based 2
  perturbation technique 2
  process 136
  variables 7, 35, 143, 154
Stress 5, 6, 13, 14, 16, 17, 25, 30, 32, 34, 41,
    42, 43, 44, 52, 56, 60, 67, 68, 69, 107,
    111, 114, 115, 117, 122, 124, 125, 128,
    130, 132
  calculation formula 34
  initial 5, 6

solution of 107, 111, 114, 117, 122, 124,
    125, 128, 130, 132
strength interference model 25, 32, 42
value and covariance of 16, 17, 43
vectors 67, 68, 69, 111
Stress-strain relationship 4, 5
    linear elastic 5
Structural 3, 23
    reliability calculation methods 23
    systems, nonlinear 3
Structural dynamics 24, 147
    nonlinear 147
Structures 1, 3, 23, 24, 52, 53, 55, 56, 59, 60,
    63, 64, 79, 80, 99, 106, 147, 148, 154,
    159, 160
    adobe masonry 3
    linear-elastic 23, 64
    static 79
System 53, 121
    failure probability 53
    mechanical propulsion 53
    nonlinear 121
    repairable 53
    reliability 53

T

Tangent stiffness method 1, 4, 99, 102, 109,
    129
Tau methodology 52
Taylor 157, 159
    fuzzy 157, 159
    interval 159
    stochastic 157, 159
Taylor expansion 1, 25, 26, 63, 64, 65, 66, 79,
    80, 90, 98, 99, 101, 102, 108, 109, 117,
    121, 127, 147, 148, 151, 155, 156, 160
    formula 25, 26, 65, 66, 90, 102, 108, 109,
    127, 151, 156
    method 1, 7, 79, 80, 98, 99, 117, 121, 147,
    148, 155
Trigonometric functions 98
Two-dimensional local averaging technique
    134

U

Uncertain coefficients 2
Uncertainty 3, 24, 53, 63, 64, 79, 98, 99, 120,
    148
    geometric 98
    multivariate interval 99
    parametric 79

V

Variability response functions 2
Variance of stress 18, 35, 42, 52, 56, 57, 59,
    60
Vector norm 71
Vibration 147, 148
    analysis 148
    equation 147

W

Weighted integral method 134

Y

Young's modulus 7, 107, 111, 114, 115, 117,
    122, 124, 125, 128, 130, 132

    interval variables of 107, 111, 114, 115,
    117, 122, 124, 125, 128, 130, 132

www.ingramcontent.com/pod-product-compliance
Lightning Source LLC
Chambersburg PA
CBHW041706210326
41598CB00007B/552